Lab-At-Home Experiments in Chemistry and Chemical Engineering

With the rise of remote and hybrid learning, there is an increasing need for flexibility around laboratory-based coursework. This book offers readers a guide to conducting a wide variety of chemistry and chemical engineering labs and experimental procedures that can be completed easily and safely outside of a traditional lab setting, including at home. It helps students and interested readers achieve hands-on learning of chemistry- and chemical engineering-based concepts without the need for sophisticated lab equipment.

- Features a comprehensive range of labs on such topics as separation processes, CO_2 capture, 3D printing, reaction kinetics, and fuel cells
- Includes detailed information on how to perform experiments at home via virtual experimental procedures and accompanying videos
- Uses household and readily accessible materials and provides safe material handling and disposal methods
- Offers solution spreadsheets where experimental results can be computed and validated
- Gives background information and relevant theory about each experiment and includes exercises at the end of each chapter
- Provides Python code and instructions on how to perform experiments, including a bowling example

This practical and useful guide is aimed at chemistry and chemical engineering students as well as general readers interested in running experiments in these disciplines. Video supplements to support comprehension of each lab and a solutions spreadsheet for Chapter 1 are available for download.

Jude A. Okolie is an Assistant Professor at Gallogly College of Engineering, University of Oklahoma. He was a postdoctoral fellow in the Department of Chemical and Biological Engineering at the University of Saskatchewan, Canada. He earned a doctoral degree at the University of Saskatchewan.

Ugochukwu Patrick Okoye earned a PhD in chemical engineering at Universiti Sains Malaysia. He worked as a postdoctoral fellow at Shenyang University of Technology, China. He is an Associate Professor at the Instituto de Energias Renovables, Universidad Nacional Autonóma de Mexico.

Lab-At-Home Experiments in Chemistry and Chemical Engineering

Jude A. Okolie and
Ugochukwu Patrick Okoye

With Contributions by Maya Fetzer

CRC Press
Taylor & Francis Group
Boca Raton London New York

CRC Press is an imprint of the
Taylor & Francis Group, an **informa** business

Designed cover image: © 2026 Shutterstock

First edition published 2026
by CRC Press
2385 NW Executive Center Drive, Suite 320, Boca Raton FL 33431

and by CRC Press
4 Park Square, Milton Park, Abingdon, Oxon, OX14 4RN

CRC Press is an imprint of Taylor & Francis Group, LLC

© 2026 Jude A. Okolie and Ugochukwu Patrick Okoye

ISBN: 978-1-032-44918-0 (hbk)
ISBN: 978-1-032-44914-2 (pbk)
ISBN: 978-1-003-37450-3 (ebk)

DOI: 10.1201/9781003374503

Typeset in Palatino
by codeMantra

Access the Support Material: https://www.routledge.com/9781032449180

Contents

Preface

The COVID-19 pandemic has compelled universities worldwide to reevaluate their approaches to teaching and course delivery. While some institutions have fully adopted remote learning, others are gradually transitioning to hybrid models that blend online and in-person instruction. However, conducting in-person laboratory experiments remains a significant challenge due to safety concerns and pandemic-related restrictions. This has sparked an ongoing debate: should universities prepare laboratory classes for future in-person delivery, or should they pivot toward developing virtual or at-home experiments? While virtual experiments offer a practical solution during times of restricted access, concerns remain regarding their impact on the quality of education and student motivation.

To address these challenges and enhance the learning experience, we have developed a set of "lab-at-home" toolkits and experimental procedures that cover a broad range of topics. These experiments include adsorption of pollutants, renewable hydrogen production, 3D printing for reactor design, data-driven machine learning processes, household pathogen removal, and the design of supercapacitors. Each procedure is designed to be simple to follow, and a standardized data reporting table is provided in each chapter to support consistent result documentation. Additionally, an introductory overview is included for every topic to provide context and learning objectives.

The book consists of eight chapters, each addressing a timely and relevant topic to help students apply theoretical knowledge in a practical, hands-on manner:

- Chapter 1 introduces an at-home experiment focused on adsorption. Students learn how to perform calculations related to adsorption equilibrium, kinetics, isotherms, and thermodynamic parameters. This experiment provides foundational knowledge for understanding pollutant removal processes, which are crucial in environmental engineering.

- Chapter 2 presents a home-based bioethanol production experiment. Bioethanol is selected due to its relevance as a renewable and clean alternative to fossil fuels. The experiment demonstrates basic fermentation principles and links them to broader discussions on sustainable energy.

- Chapter 3 offers a hands-on guide to designing 3D bioreactors. Understanding and applying 3D printing in reaction engineering is essential to bridge the gap between lab-scale designs and industrial applications.

- Chapter 4 engages students in a data collection and machine learning activity using a mini-bowling game. This chapter introduces artificial intelligence tools such as LangChain to demonstrate how structured data can be used to build predictive models and optimize performance strategies. It emphasizes the role of data science in modern engineering.

- Chapter 5 explores hydrogen production through renewable pathways, providing insights into a clean energy carrier that is central to global decarbonization goals.

- Chapter 6 features experiment for understanding the behavior of both real and ideal gases, reinforcing key concepts in thermodynamics and molecular behavior.

- Chapter 7 guides students through the construction of homemade supercapacitors. Supercapacitors are energy storage devices that offer high power density and rapid charge-discharge cycles, making them vital in emerging energy technologies.

- Chapter 8 presents experiments related to the removal of household pathogens. This topic is particularly relevant in the context of public health and promotes awareness of low-cost sanitation solutions.

Together, these chapters offer a practical, interdisciplinary approach to laboratory experiments in remote or hybrid learning environments. Video supplements to support comprehension of each lab and a solutions spreadsheet for Chapter 1 are available for download at https://www.routledge.com/9781032449180.

Acknowledgment

We express our gratitude to the General Directorate of Academic Personnel Affairs (DGAPA), particularly to the Program for Support of Innovation Projects and Improvement of Teaching (PAPIME), for the financial support granted to the project PE104123: "DEVELOPMENT OF 3D PRINTED CATALYTIC STRUCTURES AND MICROREACTORS TO SUPPORT THE EXPERIMENTAL AND THEORETICAL TEACHING-LEARNING OF BIOENERGY AT THE UNDERGRADUATE LEVEL" for the preparation of this manual. We also thank M.I. Estefania Duque Brito for helping in preparing Chapter 3 manual.

1

Adsorption Equilibrium, Kinetics, Isotherm, and Thermodynamics Experiments

Summary

How to determine the adsorption capacity and equilibrium is discussed step by step in this chapter. The theoretical framework of liquid phase adsorption is provided to reveal insights into adsorption kinetic models, isotherm, and thermodynamics. Methylene blue (MB) is used as a common model pollutant, and commercial activated carbon is used as an adsorbent. Experimental procedures to obtain the kinetic and thermodynamic data to describe adsorption system of MB-activated carbon are provided. The Excel sheets automated to help students in data curation and analysis are attached. The effect of pH and an experimental guide to determine the point of zero charge of an adsorbent are provided. The annex of this chapter describes a step-by-step practical guide to production of activated carbon from ligno-cellulosic biomass. This chapter is presented in such a way that instructors could use it as a demonstration tool while interested readers could also try the experiments at home.

1.1 Objectives of the Experiment

At the end of this chapter, students should be able to:

1. Understand the theoretical framework of adsorption equilibrium, kinetics, and thermodynamics and applications.
2. Carry out adsorption experiment to evaluate the efficiency of their carbon material in removal of pollutant.
3. Create a known concentration calibration curve in the ultra-violet spectrophotometer and Microsoft Excel sheet.
4. Model the adsorption kinetics, isotherms, and thermodynamics.
5. Use the Excel created for the experiment to compute their results and graphs.

DOI: 10.1201/9781003374503-1

1.2 Theoretical Framework on Adsorption, Kinetics, and Thermodynamics

Adsorption is a surface phenomenon that involves the contact and retaining of a liquid or gaseous substance called adsorbate on the surface of a solid material called adsorbent (Figure 1.1). The solid material is usually porous materials like activated carbon, zeolites, natural clays, sands, etc. For this to occur, the surface of the adsorbent must possess some form of energy that attracts the molecules of the adsorbate. There are two forms of adsorption namely physical and chemical adsorption. In the case of physical adsorption, it is rapid, and the rate-limiting step dominates the overall adsorption rate, which will determine the rate at which the pollutant is removed from the solution. The movement of this adsorbate molecule has sequential steps from bulk diffusion to external or internal diffusion, intraparticle diffusion, and finally adsorption. The interaction forces in physisorption are mainly van der Waals forces that comprises columbic attraction, hydrophobic interaction, $\pi-\pi$ interaction, and dipole interaction. If the system is experiencing a chemical reaction, the surface reaction will be slower than the diffusion step and hence, controls the removal of adsorbent. In other words, the surface reaction becomes the rate-limiting step. The forces of interaction in chemisorption are largely covalent bonding.

The adsorption experiment is usually conducted at a temperature above 30°C for a contact time with constant agitation at around 150 rpm. The adsorption of the pollutants by the adsorbent at equilibrium could be determined by Equation 1.1.

$$q_e = \frac{V\left(C_e - C_0\right)}{m} \tag{1.1}$$

FIGURE 1.1
Adsorbate adsorption on porous carbon material.

where q_e=equilibrium adsorption, V is the total volume, C_e is the concentration at equilibrium, C_0 is the initial concentration, and m is the mass of the adsorbent.

1.2.1 Adsorption Kinetics

The process of adsorption of pollutants on the surface of activated carbon is time dependent; therefore, understanding the rate of adsorption at the solid-liquid interface with respect to time is important for an efficient design of adsorption systems. This data can be used to fit the adsorption kinetic models. Hence, adsorption kinetics allows us to determine residence time to reach equilibrium, identify the adsorption rate-step, and predict the pathways of reactions. The transport of the pollutant molecules or adsorbate to the surface of the adsorbent where the binding site is located can be categorized into four; notably, bulk diffusion, film or external mass diffusion, intraparticle diffusion, and adsorption on the active site of the adsorbent. This movement is shown in Figure 1.2.

In bulk diffusion or convection of bulk, the transportation of the adsorbate molecules is instantaneously moved from the bulk solution to the boundary film layer surrounding the adsorbent. This movement is very fast especially

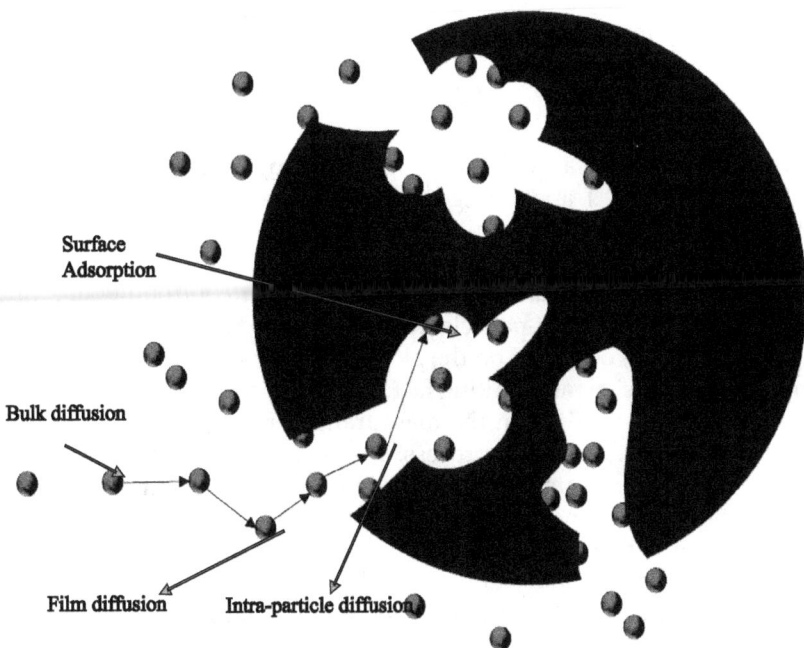

FIGURE 1.2
Adsorption stages on the activated carbon surface.

for well-stirred or mixed systems, hence, the contribution to adsorption engineering is highly negligible. In most adsorption modeling, it is usually excluded in the adsorption transport system. Another barrier to diffusion may occur during adsorption. The mass transfer resistance is associated to the bulk transport of the adsorbate molecules across the hydrodynamic boundary layer (stationary boundary layer) of the external liquid film that surrounds the adsorbent. The transport is mainly molecular diffusion governed by the concentration gradient. The diffusion rate largely depends on the hydrodynamic properties of the adsorption system.

Intraparticle diffusion, often called pore diffusion, is the internal diffusion of the molecules of the adsorbate from the external surface to the pore walls. Hence, this diffusion is independent of the adsorption system hydrodynamics, but it largely depends on the pore structure and size. The two categories of intraparticle diffusion are the pore volume diffusion and surface diffusion. The former is the molecular diffusion of solute into the fluid-filled pores, whereas the latter involves solute diffusion along the surface of the adsorbent after adsorption has occurred. Surface diffusion occurs when surface forces aren't too strong to restrict molecule mobility. This is especially significant in porous adsorbents with high surface area and narrow pores. Macropore and micropore diffusion are termed pore and surface diffusion, respectively. These diffusion types often occur simultaneously within adsorbent particles. The Weber and Morris model of Equation 1.2 is commonly used to test intraparticle diffusion [1].

$$q_t = k_i t^{\frac{1}{2}} + C_i \qquad (1.2)$$

where q_t (mg/g) is the adsorption at time t, k_i, (mg/g·min$^{0.5}$) is the rate constant, t is time (min$^{0.5}$), and C_i (mg/g), which is the intercept, provides information about boundary layer thickness.

Near representation of q_t against $\sqrt{t}$ with the line passing through the origin serves as an indicator that the predominant controlling step of the adsorption process is intra-particle diffusion. When notable deviations from the origin point are observed on the plot, it strongly suggests that pore diffusion isn't the exclusive rate-limiting factor. A slight departure from the origin can be linked to variations in the mass transfer rate during the initial and final phases of the adsorption process. The rate constant k_i for the diffusion model appears with a smaller magnitude if the mechanism at hand is the limiting stage, indicating a decrease in the collision rate of adsorbate molecules with the sorbent layers. A larger intercept value C_i, however, signifies more pronounced boundary effects, implying heightened resistance to the unrestricted diffusion of molecules. Hence, steps limiting the rate tend to have higher C_i values.

Following the transport of the adsorbate to vacant sites, an adsorption bond forms. In the context of physical adsorption, the actual physical bonding of the adsorbate to the adsorbent occurs rapidly. The slowest step among the discussed diffusion stages becomes the rate-limiting step. This step governs the overall rate of adsorbate removal from the solution. However, if the adsorption process involves a chemical reaction that alters the nature of the molecule, this chemical transformation might proceed at a slower pace than the diffusion step. Consequently, the chemical reaction could potentially control the rate of removal.

In aqueous solution, three kinetic models are popularly used, namely pseudo first-order, pseudo second-order, and Elovich models are deployed to describe the adsorption kinetic constants.

1.2.1.1 Pseudo First-Order Kinetic Model

Lagergren [2] proposed the pseudo first-order (PFO) kinetic model in 1889 following his experiments in oxalic and malonic acid adsorption on charcoal. The differential model is presented as shown in Equation 1.3:

$$\frac{dq}{dt} = k_1 \left(q_e - q \right) \tag{1.3}$$

Integrating Equation 1.5 with initial conditions $t = 0 \rightarrow q = 0$ will give Equation 1.4.

$$q_t = q_e \left(1 - e^{-k_1 t} \right) \tag{1.4}$$

The rate constant k_1 exhibits sensitivity to the dominant process conditions. Studies indicate a decrease in k_1 with higher initial bulk concentrations. This trend can be elucidated as follows: The reciprocal of $1/k_1$ explains the timespan for the process to attain equilibrium. A more extensive period is necessitated (corresponding to a smaller k_1 value) when the initial concentration C_0 is more substantial. Besides, it is possible for k_1 to have an increasing trend with C_0 or unaffected by C_0. In general, k_1 is influenced by the experimental conditions such as the temperature and solution pH.

1.2.1.2 Pseudo Second-Order Kinetic Model

Pseudo second-order kinetics (PSO) assumes that the rate-limiting step is due to chemical reaction or chemisorption, which controls the removal capacity. Hence, the rate of adsorption is independent of adsorbate concentration, but it depends on the adsorption capacity [3]. Within this mechanism, the dynamics of the sorption process are intertwined with a pair of competitive, reversible second-order reactions at high adsorbate/adsorbent ratios. Conversely, an irreversible second-order reaction prevails at lower adsorbate/adsorbent ratios [2]. A major advantage of the pseudo second-order model over the pseudo first-order model is that equilibrium adsorption capacity can be calculated from the model; hence, there is theoretically no need to determine it experimentally. The differential form of the pseudo second-order kinetic model is presented in Equation 1.5.

$$\frac{dq}{dt} = k_2 \left(q_e - q_t\right)^2 \tag{1.5}$$

Integrating Equation 1.6 with boundary conditions of $t = 0 \rightarrow q = 0$ will yield Equation 1.6.

$$\frac{t}{q_t} = \frac{1}{k_2 q_e^2} + \frac{t}{q_e} \tag{1.6}$$

where k_2 (g/mg·min) is the equilibrium rate constant of the pseudo second-order model. A plot of $\dfrac{t}{q_t}$ versus t gives a straight-line graph, where $\dfrac{1}{k_2 q_e^2}$ can be determined as the intercept and $\dfrac{1}{q_e}$ as the slope of the plot.

1.2.1.3 Elovich Model

The Elovich kinetic model is a widely used mathematical model to describe the adsorption or reaction behavior. Developed by Lev E. Elovich in the mid-20th century, this model provides insights into the mechanisms of adsorption and is especially practical in various scientific and industrial applications. It is usually adapted to describe complex reaction mechanisms involving adsorption and surface reactions. It can provide an accurate representation of the adsorption mechanisms and offers mathematical equation that relates the rate of adsorption to the extent of adsorption or contact time. The model can be used to determine adsorption capacities and kinetics, since it provides insights into adsorption-desorption of various pollutants on the surface of adsorbents. This is vital for assessing the effectiveness of adsorbents in the removal of pollutants.

The Elovich equation is recognized for its ability to characterize chemisorption while disregarding desorption, as highlighted in the work of Acevedo et al. [4]. However, it has a notable limitation as it predicts an infinite adsorption capacity (q) over extended adsorption periods. Consequently, its applicability lies more in describing kinetic behavior, particularly when desorption is negligible due to low surface coverage, rather than achieving equilibrium.

The model was proposed in 1934 by Roginsky and Zeldovich when analyzing the adsorption of CO on manganese oxide [5]. The model is expressed as given in Equation 1.7.

$$\frac{dq}{dt} = \theta e^{(-\beta q)} \tag{1.7}$$

Integrating the above equation gives Equation 1.8.

$$q_t = \frac{1}{\beta}\left(1 + \theta\beta t\right) \tag{1.8}$$

where θ (mg/g·min) is the surface coverage, and β (g/mg) is related to activation energy of chemisorption.

1.2.2 Adsorption Isotherm

There are many adsorption isotherm models and modified models based on limitations of the traditional models for task-specific applications. However, there has not been any significant shift from these traditional equilibrium isotherm models because they could be largely used to describe the adsorbate-adsorbent system and to evaluate the adsorption process. Understanding the equilibrium state within a well-defined adsorbate-adsorbent system holds vital importance for assessing the capability of adsorbing water pollutants, making informed choices about suitable adsorbents, and formulating designs for batch, flow-through, or fixed-bed adsorption systems. The equilibrium status within the considered adsorptive system depends on factors such as the affinity between the adsorbate and adsorbent, the attributes of the adsorbate, and the characteristics of the aqueous solution such as the pH, initial concentration, temperature, and more. Here three equilibrium isotherms are considered, namely Langmuir, Freundlich, and Temkin isotherms.

1.2.2.1 Langmuir Isotherm

The Langmuir model provides a precise description of how a monolayer of adsorbate forms on the external surface of the adsorbent, with no additional adsorption occurring thereafter. The isotherm model was derived based on the following assumptions:

- Adsorbate molecules are adsorbed at discrete points on the adsorbent, which are referred to as adsorption sites. Each site is occupied by a single molecule of the adsorbate and no further adsorption occurs.
- The energy of an adsorbed species remains constant across the entire surface, regardless of whether nearby adsorbed molecules are present or absent. Hence, the forces between adjacent adsorbed molecules are assumed to be sufficiently weak, and the likelihood of adsorption onto an empty site is not influenced by the occupancy level of active centers on the adsorbent. Besides, this assumption implies that the surface is energetically uniform and exhibits consistent adsorption activity throughout.

- The maximum amount of adsorption on the surface of adsorption relates to a monolayer adsorption.
- Adsorption occurs through the collision of adsorbates with vacant sites, with no interaction between adsorbed molecules on adjacent sites.
- The rate of desorption of adsorbed species depends on the amount adsorbed molecules.

The Langmuir isotherm can be represented by Equation 1.9 [6].

$$q_e = \frac{q_{max} K_L C_e}{1 + K_L C_e} \qquad (1.9)$$

where q_e (mg/g) is the amount of the adsorbate adsorbed per unit mass of the adsorbent at equilibrium, C_e (mg/L) is the concentration of the adsorbate in the solution at equilibrium, q_{max} is the maximum monolayer adsorption capacity, and K_L (L/mg) is the Langmuir constant that relates to adsorbate-adsorbent interactions and adsorption energy.

1.2.2.2 Freundlich Isotherm

Freundlich isotherm arguments the limitations of Langmuir isotherm, considering that possible interactions in fact may exist between the adjacent adsorbed species, and the adsorbent surface may be heterogeneous. So, it is an empirical model used to elucidate multilayer adsorption with interaction of the adsorbates. The assumption is that the initially occupied sites are the ones with stronger binding, and as adsorption progresses along the adsorbent's centers, the binding energy exponentially diminishes toward the end. So, the model appropriately describes the adsorption on heterogeneous surfaces with different affinities (binding energy). The Freundlich model is presented in Equation 1.10 [7].

$$q_e = K_F C_e^{\frac{1}{n}} \qquad (1.10)$$

K_F $\left(\text{mg/g·} (\text{L/min}) 1/n \right)$ is the Freundlich adsorption capacity, n is the adsorption intensity, and the degree of nonlinearity between the concentration of the solution and adsorption capacity could be determined from the relation $0 < 1/n < 1$, where $1/n = 1$ indicates linear adsorption process and $1/n > 1$ reveals collective adsorption.

1.2.2.3 Temkin Isotherm

Experimental data derived for both Langmuir and Freundlich is used to elucidate the Temkin isotherm. Temkin isotherm was derived assuming that the adsorption heat of all molecules in a layer decrease linearly as the surface coverage increases. Hence, the adsorption process has a character-istic of uniformly distributed binding energy, which could be used to learn the adsorbate-adsorbent interaction. Equation 1.11 expresses the Temkin iso-therm model [8].

$$\text{Temkin}: q_e = \frac{RT}{B_T} \ln(A_T C_e) \tag{1.11}$$

where R (8.314 J/mol·K) is the universal gas constant, T is the temperature in Kelvin, B_T (kJ/mol) is a constant related to the adsorption heat, and A_T. is the equilibrium binding constant. A plot of q_e versus $\ln C_e$ can be used to deter-mine the slope and intercept, RT/B_T and $RT \ln A_T/B_T$, respectively.

1.3 Thermodynamics

The analysis of the adsorption thermodynamics will help the student to know if the adsorption is feasible and spontaneous or not. The enthalpy (ΔH), entropy (ΔS), and Gibbs free energy (ΔG) could be calculated from the experimental data. The Gibbs free energy is calculated from Equation 1.12, whereas the classical Van't Hoff equation can be used to calculate the values of the enthalpy and entropy as shown in Equation 1.13.

$$\Delta G = -RT \ln K_c \tag{1.12}$$

$$\ln K_c = \frac{\Delta S}{R} - \frac{\Delta H}{RT} \tag{1.13}$$

where $K_c = q_e/C_e$, K_c is the distribution coefficient, R (8.314 J/mol·K) is the ideal gas constant, and T is the adsorption temperature in Kelvin.

1.4 Experimental Procedure for Obtaining Adsorption Equilibrium, Isotherm, and Kinetic Data

1.4.1 Goals of the Study

Prepare solutions starting with a known pollutant, in this case MB.

- Perform a serial dilution as described in the previous section.
- Use the spectrophotometer to measure the absorbance of solutions.

- Generate a standard curve and use the standard curve to determine the equilibrium concentration of a solution.
- Perform adsorption experiments to determine the adsorption capacity, isotherm, kinetics, and thermodynamics of adsorption.

1.4.2 Student Learning Outcomes

- Students upon completion of this lab experiment will be able to:
- Determine the mass of solute needed to make at %(w/v) solution.
- Make a stock solution of the appropriate concentration.
- Create a series of solutions of decreasing concentrations via serial dilutions.
- Use the spectrophotometer to measure the absorbance of a solution.
- Use Excel and make a standard curve, fit the points to a linear model, and use the R^2 value to determine the quality of the calibration curve.
- Use the standard curve to calculate the concentration of a solution.

1.4.3 Materials, Chemicals, and Equipment

1. The MB solutions were prepared using distilled water. The experiments for adsorption were done in batch method.
2. Commercial activated carbon.
3. MB—CAS # 122965-43-9.
4. 10 mL (5–8 of them are good) and 250 mL volumetric flasks.
5. Erlenmeyer flasks (50–100 mL) with glass stoppers or plastic bottles with caps.
6. Small rectangular stainless bath to serve as water bath and a thermometer (you need not more than 40°). The stirring could be done by intermittent swirling of the flasks.
7. A ultraviolet-visible (UV-Vis) spectrophotometer to analyze the removal of the pollutants.
8. Cuvette of 1 cm path length, which is the liquid sample holder used for analysis in UV-Vis light spectrophotometer.
9. Please wear your personal protective equipment (PPE), a lab coat, gloves, and transparent glasses for your safety.

The MB solutions were prepared using distilled water. The experiments for adsorption were done in batch method.

1.4.4 Creating a Calibration Curve

The first step to determine the adsorption equilibrium is making a calibration curve in a UV-Vis spectrophotometer. So, if you have an idea of the pollutant you are working with, you can make a calibration curve using a known concentration of the pollutant. The concentration versus absorbance curve can be determined with a portable UV-Vis. The UV-Vis spectrophotometer is somewhat affordable and can be bought in online stores. Besides, the students should be able to see visually the difference in color before and after the adsorption experiment for most colored dyes if an adsorption ever occurred.

To make a calibration curve, known concentration of the adsorbate or pollutant is diluted with distilled water (dilution factor should be noted because it is important in the calculation of the equilibrium concentration).

a. Label 6 (points could be more, but ensure it covers the whole range of the concentration) clean 10 mL volumetric flasks for your standard solutions.

b. Prepare a stock solution of 50 mg/L in a 250 mL volumetric flask.

c. Using a serial dilution, create 6 standards from 1 to 6 mg/L MB from the stock solution.

The calculation to perform serial dilution is given in Equation 1.14.

$$C_1 \times V_1 = C_2 \times V_2 \tag{1.14}$$

where C_1 is the concentration of the stock solution, V_1 is the volume of the stock solution (volume of the flask), C_2 is the final concentration, and V_2 is the final volume. Figure 1.3 shows the serial dilution in six volumetric flasks.

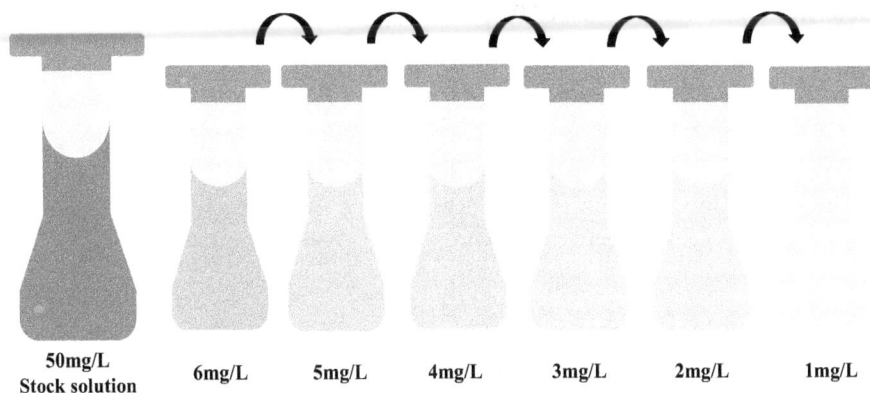

| 50mg/L Stock solution | 6mg/L | 5mg/L | 4mg/L | 3mg/L | 2mg/L | 1mg/L |

FIGURE 1.3
Serial dilution from stock solution using 10 mL volumetric flasks.

d. Determine the maximum wavelength λ_{max} of the solution using a stock solution. This will be the wavelength (λ_{max}) that corresponds to the maximum absorbance on the spectrum. The wavelength is measured in nanometers (nm), since they are very small.

e. Measure the length of the cuvette, the measurement is in centimeter, and this is called the path length L (cm).

f. Take the diluted concentration in 6 mg/L using a pipette and put it in a 1-cm quartz cuvette.

g. Make sure to dry the outside of the cuvette with a clean napkin, and then insert it in the UV-Vis spectrophotometer. Also, with another cuvette, add distilled water and insert in the other sample holder of the UV-Vis spectrophotometer.

h. Run the analysis to record the absorbance and do this for all the samples.

   Note: Spectrophotometer is used for quantitative analysis to determine concentration based on the principle of light absorption. In this equipment, a beam of light with specific wavelength passes through a solution; some of the light spectrum is absorbed by the solution, and the remaining light that comes out could be used to calculate the absorbance of that solution. Hence, an absorbance versus the concentration plot, which is automatically obtained via the UV-probe software, is the calibration curve.

i. Plot the absorbance versus the concentrations (i.e., absorbance as y-axis and concentration as x-axis).

j. The plot should be a straight line that fits the points, which confirms the proportionality of concentration with absorbance.

A high-quality calibration curve for adsorption should have a coefficient of determination R^2 of at least 0.99.

1.4.5 Theory

To quantitatively determine the concentration at any time interval, a single component analysis using Beer-Lambert's law, which shows the linear relationship between the absorbance and the concentration, can be used. Beer-Lambert's law is generally expressed as given in Equation 1.15.

$$A_i = \varepsilon_\lambda \cdot L \cdot C \qquad (1.15)$$

A_i is the absorbance of the i component, ε_λ is the molar absorptivity at wavelength λ in nm, L is the path length of the quartz cuvette (usually 1 cm or specified by the vendor), and C is the concentration (mg/L). Figure 1.4 is a graph of the absorbance at different concentrations from 1 to 6 mg/L with

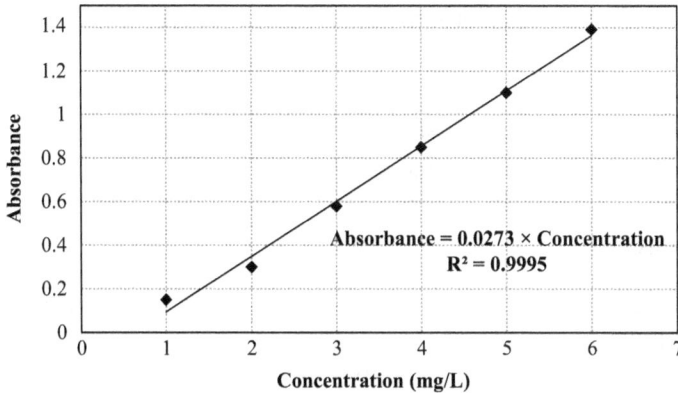

FIGURE 1.4
Sample calibration curve using random data to demonstrate Beer Lambert's law.

R^2 of 0.9995. The students should be able to obtain a similar type of graph (Figure 1.4) with concentrations from 1 to 6 mg/L.

The slope of the straight line is the molar absorptivity ε_λ (L/mol/cm), assuming the path length of the cuvette is 1 cm. The above graph shows that the molar absorptivity is 0.0273. Manually, the molar absorptivity can be determined from the graph. After plotting the straight-line graph, the slope can be determined between the two points of x- and y-axis (Equation 1.16). The slope is then divided by the pathlength of the cuvette to obtain the molar absorptivity (Equation 1.17).

$$\text{slope} = \frac{\left(y_2 - y_1\right)}{\left(x_2 - x_1\right)} \tag{1.16}$$

$$\varepsilon_\lambda = \frac{\text{slope}}{\text{pathlength}\,(\text{cm})} \tag{1.17}$$

1.5 Determining the Adsorption Equilibrium, Isotherm, Kinetics, and Thermodynamics

1.5.1 Learning Outcomes

At the end of this section, the students should be able to:

- Perform adsorption experiments using different concentrations of MB.

- Determine the equilibrium concentration, which is the concentration when the removal capacity is constant.
- Use the Excel sheet to calculate the adsorption capacity at equilibrium.
- Perform the adsorption kinetics and thermodynamic experiments.
- Use the Excel to compute the isotherm, kinetics, and thermodynamics.

1.5.2 Safety

- Students should always wear goggles and personal protective equipment.
- Do not dispose of this dye in the sink; make sure it is properly disposed of in a waste container that is clearly labeled.
- Note that MB is a dye and will stain things, so be careful to use only glassware.
- Do not dispose of the activated carbon in the sink.
- Wash all the glassware to remove the dyes after the experiment.

1.5.3 Experimental Steps

- Students should prepare 50 mg/L of stock solution.
- Dilute the stock solution to obtain 10, 8, 6, and 4 mg/L concentrations in four volumetric flasks.
- Clearly label the volumetric flasks. If there are no volumetric flasks, students could use a calibrated clean bottle.
- Pour some water in a rectangular bowl made of stainless steel and of depth enough to allow your samples to immerse up to the volume of the MB solution.
- In the lab, you can use a thermostatic water bath shaker with time, temperature, and agitation speed controller (Figure 1.5). You can set your parameters using the manual of the water bath.

1.5.4 Running the Adsorption Experiment

- Take about 20 mL of the diluted concentrated solutions and place them in a flask with a stopper.
- Allow the solution to equilibrate for 15 minutes.
  Note: The first experiment to determine the adsorption equilibrium will be conducted at 25°C (room temperature), which is assumed to be the temperature of your water. If it's lower, then the water must be heated, and the temperature can be adjusted with intermittent

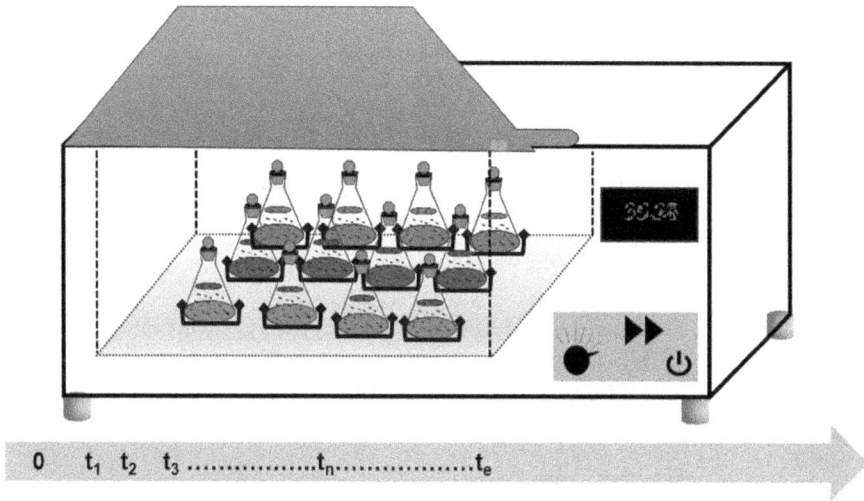

FIGURE 1.5
Adsorption experimental setup in a thermostatic water bath shaker.

heating. The temperature can be monitored with digital thermometer. This will apply for experiments carried out at 40°C and 50°C.

Any alternative heater with mixer could be used at home to perform this experiment in place of a water bath shaker.

- Add a magnetic bar inside the flask, and add the commercial activated carbon of 50–200 mg. The range of mass of the activated carbon is selected for the students to investigate the effect of activated carbon dosage.

- Start the experiment and adjust the temperature, while monitoring with digital thermometer (allowable temperature of thermometer up to 100°C).

- Take samples of the solution at different time intervals and measure their residual concentration. Note: Always allow the carbon to settle before analysis with UV-Vis spectrophotometer, otherwise, the carbon might absorb some light, resulting in poor quality of the result.

- Note the absorbance at which the residual concentration does not change. This concentration is the equilibrium concentration, and the calculated adsorption capacity or uptake at this concentration is the adsorption equilibrium.

Record your data for each temperature 25°C, 40°C, and 50°C as shown in Table 1.1. Students can use the Excel sheet in the annex to automatically analyze their results and obtain the plots if they have the data of their experiments.

TABLE 1.1

Sample Sheet for Computation of Adsorption Data

Temperature (°C or °F)					
Sample (e.g., 10 mg/L)	Measurement	Time (minutes)	Absorbance	Dilution Factor	Concentration (mg/L)
	1				
	2				
	3				
	.				
	.				

1.6 Determining the pH of the Solution (This Should Not Be Done at Home or Care Should Be Taken While Handling Acids)

In adsorption, the solution pH plays an important role in the removal of the pollutants. It affects both biological and chemical processes in water. Also, it could facilitate the precipitation of heavy metals in wastewater. Conducting adsorption experiments in the appropriate pH region facilitates the pollutant removal from water and wastewaters.

Safety: Wear your protective gloves and goggles and nose mask (HCl is fuming and could cause respiratory problems). DO NOT ADD WATER TO ACID to avoid violent splashing.

Take six flasks and clearly label them pH 2, 4, 6, 8, 10, and 12.

- Prepare 0.1 M solution of sodium hydroxide and hydrochloric acid in two volumetric flasks.
- Transfer 10 mL of 5 mg/L solution that you prepared into the six flasks.
- Add 10 mg of activated carbon into the six flasks.
- Adjust the pH from 2 to 12 using the prepared solutions of 0.1 M NaOH and 0.1 M HCl.
- Place the experiment in a water bath and agitate for 24 hours at 30°C.
- Determine the final pH using the benchtop scientific potentiometer. Also, the final concentration can be determined using the UV-Vis spectrophotometer.

- The point of zero charge pH_{pzc}, which is the point at which the adsorbent surface charge is neutral, could be obtained from the intersection of the plot of the initial pH_i and final pH_f, i.e., $pH_i = pH_f$.

References

1. W.J. Weber and C. Morris, "Kinetics of adsorption on carbon from solution," *Journal of the Sanitary Engineering Division, American Society of Civil Engineers,* vol. 89, pp. 31–60, 1963.
2. S. Lagergren, "About the theory of so-called adsorption of soluble substances," *Vetenskapsakademies Handl,* pp. 1–39, 1898.
3. Y.S. Ho and G. McKay, "Pseudo-second order model for sorption processes," *Process Biochemistry,* vol. 34, pp. 451–465, 1999, https://doi.org/10.1016/S0032-9592(98)00112-5.
4. B. Acevedo and C. Barriocanal, "Simultaneous adsorption of Cd^{2+} and reactive dye on mesoporous nanocarbons," *RSC Advances,* vol. 5, pp. 95247–95255, 2015, https://doi.org/10.1039/C5RA21493A.
5. S.Z. Roginsky and J. Zeldovich, "An equation for the kinetics of activated adsorption," *Acta Physicochimca (USSR),* pp. 554–559, 1934.
6. S. Langergen and B.K. Svenska, "Zur Theorie der sogenannten Adsorption gelöster Stoffe," *Zeitschrift Für Chemie Und Industrie Der Kolloide,* vol. 2, p. 15, 1907, https://doi.org/10.1007/BF01501332.
7. H.M.F. Freundlich, "Über die adsorption in losungen (adsorption in solution)," *Zeitschrift Für Physikalische Chemie,* vol. 57, pp. 385–490, 1906.
8. M.J. Temkin and V. Pyzhev, "Recent modifications to Langmuir isotherms," *Acta Physiochimca URSS,* vol. 12, pp. 217–225, 1940.

Annex: Procedure to Prepare Activated Carbon from Biomass by Chemical Activation

Things to consider before choosing biomass:

- Availability
- Abundance
- Fixed carbon content of the biomass (this is important to have appreciable yield)

Procedure:

- Wash your biomass to remove impurities such as sand, metals, and any debris.
- Dry the biomass in the oven until the moisture content is less than 1%.
- Reduce the particle size by grinding the biomass with a blender or industrial grinder.
- Soak your biomass with an activating agent (these agents are salts or acids) in a 100 mL beaker. For example, these agents could be KOH, NaOH, K_2CO_3, Na_2CO_3, H_2SO_4, H_3PO_4, etc. Ensure that the biomass is entirely soaked in the aqueous solution containing your activating agent.

  Note: Activating agent removes impurities (inorganics) from the biomass, opens the pores and produce very high ordered surface area. These activating agents have temperature standard to realize optimum reaction with biomass and depending on the type of activating agent, the nature and type of the pores could vary widely.
- You can vary the weight ratios (wt./wt.) of the agent and the biomass to obtain the optimal amount of the activating agent.
- Stir the slurry, i.e., aqueous solution of the biomass and activating agent and heat the solution at 85°C for at least 4 hours. After this, evaporate the excess water in the oven at 105°C.
- Put the dried biomass gent in a crucible and place it in a furnace.
- Connect the furnace to a nitrogen or argon tank and initially increase the inert gas flow to ensure inert atmosphere. Then reduce the gas flow rate to100 mL/min.
- Activate your biomass by varying the temperatures from 500°C to 900°C for 1–2 hours.
- After activation, perform acid washing by adding 1 M HCl to the biomass in a 100 mL beaker to remove ashes, excess of the alkaline salts and clean the carbon pores. It is important to note the pH.
- After acid washing, wash the activated carbon with distilled water using the vacuum filtration equipment (a filtration flask connected to a vacuum pump). Monitor the pH and stop when the pH reaches 6.5–7.
- Dry the washed activated carbon in an oven at 105°C and store in an airtight container (Table 1.2).

TABLE 1.2

Activated Carbon Production Influencing Factors

Factors	Low	High	Yield (%)
Activation temperature (°C)	500	900	
Activation time (minutes)	60	120	
Biomass: Agent ratio (wt./wt.)	3	1	

Note: For physical activation, the biomass is first carbonized from 500°C to 800°C and then activated by passing CO_2 or steam through the carbon bed at >800°C for 1–2 hours. The main advantage of physical activation is that chemicals that may become a source of pollution if not properly disposed are not required.

Yield of biomass could be calculated using Equation A.1 below:

$$\text{Yield } (\%) = \frac{\text{Mass of initial biomass } (g)}{\text{Mass of activated carbon } (g)} \times 100 \qquad \text{(A.1)}$$

2

Bioethanol Production at Home

2.1 Background on Ethanol Production

Ethanol, a biofuel, can be produced through the fermentation of sugars by yeast (*Saccharomyces cerevisiae*). This process is widely used in industrial and home settings. Ethanol production through alcoholic fermentation is a process that has been practiced for centuries, serving as the foundation for industries such as brewing, biofuel production, and even pharmaceuticals. At its core, fermentation is a metabolic pathway where microorganisms, primarily yeast, convert sugars into ethanol and carbon dioxide under anaerobic conditions. Understanding this process is crucial not only for appreciating its historical and industrial significance but also for recognizing its role in sustainable energy solutions and waste-to-resource technologies. By studying fermentation, learners gain insight into biochemical reactions, energy transformation, and the practical application of microbiology in everyday life.

Almost all ethanol is produced from the fermentation of corn glucose or sucrose, however, the issue of food versus fuel competition has stimulated interest in alternative feedstocks for ethanol production [1]. Over the past two decades, advancements in ethanol production from non-food plant sources have progressed significantly, making large-scale production from these alternative sources increasingly viable. As a result, agricultural residues such as corn stover (including cobs and stalks), sugarcane bagasse, wheat or rice straw, forestry and paper mill waste, the paper fraction of municipal solid waste, and dedicated energy crops which are collectively referred to as "biomass", can all be transformed into fuel ethanol. Despite the remarkable progress in bioethanol production technologies, several challenges remain that require further research and innovation. A deeper understanding of the fermentation processes involved in ethanol production is essential to improve efficiency, reduce costs, and make bioethanol a more competitive and sustainable alternative to fossil fuels.

Most agricultural biomass rich in starch can serve as a viable substrate for ethanol production through microbial fermentation. Common examples of these starchy feedstocks include corn (maize), wheat, oats, rice, potatoes, and cassava. On a dry weight basis, grains such as corn, wheat, and sorghum (milo) contain approximately 60%–75% starch, which can be hydrolyzed

DOI: 10.1201/9781003374503-2

into hexose sugars with a slight increase in mass (the stoichiometric ratio of starch to hexose is 9:10), making them excellent resources for fermentation processes [2].

Compared to the direct fermentation of simple sugars, fermenting starch-based materials is more complex, as starch must first be broken down into fermentable sugars before it can be converted into ethanol. Initially, starch is hydrolyzed using α-amylase to prevent gelatinization, followed by high-temperature cooking (140°C–180°C) [3]. The liquefied starch is then further hydrolyzed into glucose by glucoamylase. The resulting glucose, or dextrose, is fermented by microorganisms to produce ethanol and carbon dioxide as a byproduct. Industrial-scale ethanol production from starchy materials commonly relies on high-temperature cooking, which enhances the efficiency of starch saccharification and ensures high ethanol yields through effective sterilization of unwanted microbes. However, this method is energy-intensive and costly, largely due to the high temperatures required and the significant quantities of amylolytic enzymes needed. To address these challenges, alternative approaches such as non-cooking and low-temperature fermentation systems have been developed to reduce production costs [1]. However, understanding these fermentation processes presents several challenges as there are complex mechanisms and pathways involved. Moreover, teaching them is also challenging especially in a laboratory setting. Therein lies the motivation of developing a lab at home experimental procedure to explain ethanol production.

Conducting simple ethanol production experiments at home offers an accessible and engaging way to explore the principles of fermentation. Through hands-on practice, students can observe how factors such as temperature, sugar concentration, and yeast type influence the efficiency of ethanol production. These experiments not only make abstract biochemical concepts more tangible but also promote scientific inquiry by encouraging learners to formulate hypotheses, collect data, and analyze results. Furthermore, this approach helps build foundational skills in experimental design and process optimization, which are essential in both academic research and industrial applications of bioethanol production.

Understanding the key parameters and techniques is crucial for successful ethanol production.

2.1.1 Objectives

- The main objectives of this experiment are:
- To activate *Saccharomyces cerevisiae* for optimal fermentation.
- To conduct alcoholic fermentation under controlled conditions.
- To quantify glucose consumption and ethanol production using analytical techniques.
- To analyze the fermentation kinetics and efficiency.

2.2 Materials and Methods

2.2.1 Yeast Activation

2.2.1.1 Preparation of Yeast Inoculum

Saccharomyces cerevisiae is activated by inoculating 1.5% (w/v) of yeast in a liquid culture medium. This pre-culture is essential for ensuring a high concentration of active yeast cells at the start of the fermentation process. The high concentration ensures a dominant yeast strain and reduces the risk of contamination.

2.2.1.1.1 *Culture Medium Composition*

The liquid medium contains the following components (per liter):

Glucose $=10\,g$

Yeast extract $=4\,g$

$KH_2PO_4=2\,g$

$(NH_4)_2SO_4=3\,g$

$MgSO_4 \cdot 7H_2O=1\,g$

Casein peptone $=0.4\,g$

Gentamicin (or broad-spectrum antibacterial): $10\,ppm$ (to prevent bacterial growth).

Since the experiment is supposedly to be performed at home, Table 2.1 outlines alternatives that could be used for the experiment.

TABLE 2.1

List of Materials and Alternatives Required for the Ethanol Production Experiments

Component	Purpose	Can It Be Used at Home?	Home-Friendly Alternative
Glucose (10 g)	Carbon source (main substrate for fermentation)	Yes, with alternatives	Table sugar, honey, or corn syrup
Yeast extract (4 g)	Nitrogen, vitamins, and growth factors	Not commonly available at home	Nutritional yeast flakes or marmite/vegemite
KH_2PO_4 (2 g)	Buffering agent and phosphorus source	Rarely at home	Baking powder or cream of tartar (partial substitute)
$(NH_4)_2SO_4$ (3 g)	Nitrogen source	Not typically at home	Use garden fertilizer carefully or rely on yeast nutrients
$MgSO_4 \cdot 7H_2O$ (1 g)	Magnesium source (enzyme cofactor)	Rarely at home	Epsom salt (magnesium sulfate)
Casein peptone (0.4 g)	Peptides/amino acids for growth	Not at home	Milk, skim milk powder, or whey protein
Gentamicin (10 ppm)	Prevent bacterial contamination	Not available at home	Perform experiment without it, focus on cleanliness and sterilization

2.2.1.1.2 Incubation Conditions

2.2.1.1.2.1 If Performed in a Laboratory The yeast is incubated for 24 hours at 30°C with constant agitation at 150 rpm. These conditions promote rapid yeast growth and activation.

2.2.1.1.2.2 At Home

- **Incubation Time:** Keep 24 hours—easy to maintain.
- **Temperature (~30°C):**
  - **Warm Spot in the House:** Place the container near a warm appliance (like on top of a refrigerator), or in a cupboard where it stays warm.
  - **Sunlight:** During the day, place near indirect sunlight (not direct heat to avoid overheating).
  - **Warm Water Bath:** Place the fermentation container inside a larger container filled with warm water (~30°C). Refresh the warm water every few hours.
  - **Use an Oven with the Light On:** Some ovens maintain a low temperature with just the light bulb on (check with thermometer first).
- **Agitation (Constant Shaking at 150 rpm):**
  - **Manual Stirring:** Stir the mixture every few hours with a clean spoon to mimic agitation.
  - **DIY Shaker:** Place the container on a washing machine or massage device on a low, continuous setting—this works surprisingly well!
  - **Swirling:** Give the container a gentle swirl several times during the day.
- **Container Suggestion:** Use a tightly sealed jar or bottle, but not airtight—gas (CO_2) needs to escape. Use plastic wrap with small holes or a balloon with a pinprick hole over the top to allow gas release.

2.2.2 Alcoholic Fermentation

2.2.2.1 Fermentation Setup

The fermentation setup presented is a simplified, home-friendly version of a typical lab-scale anaerobic fermentation system, designed to safely and effectively carry out alcoholic fermentation.

2.2.2.1.1 Fermentation Vessel (Glass Bottle or Jar)

- In this setup, a **120 mL glass bottle** is recommended, with a working volume of approximately **80 mL** of fermentation medium. If a laboratory serological bottle is not available, any clean, heat-resistant glass bottle or jar with a tight-sealing lid can be used.

- It's essential that the container is well-cleaned and sterilized (boiled or cleaned with food-safe sanitizer) to minimize contamination.
- The working volume is deliberately less than the total bottle capacity to allow headspace for gas expansion and CO_2 buildup during fermentation, preventing overflow.

2.2.2.1.2 Fermentation Medium and Inoculation

- The liquid inside contains the glucose-based fermentation medium prepared earlier (homemade sugar solutions with yeast nutrients). Yeast, such as *Saccharomyces cerevisiae*, is introduced into this medium to initiate fermentation.
- The previously reactivated yeast is inoculated at a concentration of 10% (v/v). This ensures a high initial yeast population for efficient fermentation. The high concentration reduces the risk of contamination by wild yeasts or bacteria (e.g., lactic acid bacteria) that could produce off-flavors or reduce ethanol yield. The primary function of yeast is to metabolize fermentable sugars. This process, known as glycolysis, breaks down sugars into pyruvate, which is then converted into ethanol and CO_2 in the absence of oxygen.

2.2.2.1.3 Sealed Cap with Airlock System

- The bottle is sealed with a cap that has been adapted to fit an S-shaped airlock.
- The airlock is crucial for maintaining anaerobic conditions—meaning no oxygen enters the bottle, which is necessary for proper alcoholic fermentation.
- As yeast ferments the sugars, carbon dioxide (CO_2) is produced as a by-product. The S-shaped airlock allows this gas to escape safely without allowing ambient air (and potential contaminants like oxygen or unwanted microbes) to enter the system.

2.2.2.1.4 Liquid Barrier in the Airlock

- The airlock contains a small amount of liquid (often alcohol like ethanol, or clean water with a bit of sanitizer).
- This liquid barrier acts as a gas trap: it lets CO_2 bubbles escape while blocking external air from entering.
- Using alcohol in the airlock instead of water adds an extra layer of microbial protection, preventing contamination from airborne bacteria or fungi.

2.2.2.1.5 Positioning the Setup

- Place the setup in a warm location (around 28°C–32°C) to promote active yeast metabolism. Options include warm water baths, near appliances, or inside a cupboard.
- Ensure the vessel is placed on a stable surface, and avoid shaking excessively once fermentation begins to prevent clogging the airlock.

2.2.2.1.6 Observation and Safety

- You can observe fermentation progress by watching for bubbling in the airlock, which indicates active CO_2 production.
- Since the vessel is sealed, the system contains the gas safely, and the airlock prevents pressure buildup.
- Always ensure the airlock liquid level remains adequate to maintain the barrier.

2.2.3 pH Adjustment

The final pH is adjusted to 5 using buffers or 5 N HCl or NaOH. Maintaining the optimal pH is crucial for yeast activity and ethanol production. Yeast, such as *Saccharomyces cerevisiae*, thrives within a specific pH range, typically 4.0–6.0, with an ideal around 4.5–5.5 depending on the strain. Too high (pH ≥7) or too low (pH ≤3.5) of the pH stresses the yeast and results in slow fermentation or death of the yeast (Figure 2.1).

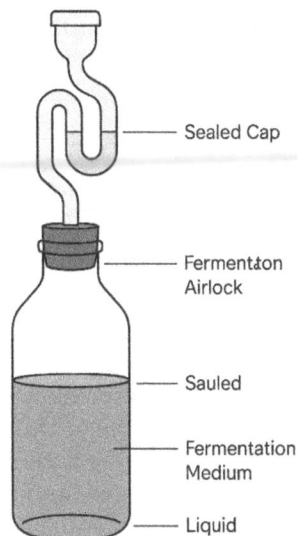

FIGURE 2.1
Schematics of the fermentation setup.

TABLE 2.2

Table for Documenting Experimental Results

Temp. (°C)	Time (hours)	pH	Inoculum Ratio (w/v%)	%Eth
20	24	3.5	5	
30	48	4.5	10	
35	72	5.5	15	
40	96	6.5	20	

2.2.4 Incubation

The serological bottles are sealed with butyl rubber stoppers and aluminum seals to create an anaerobic environment. They are then incubated at 30°C. You can achieve this temperature at home in a dark and warm place, usually near our cooking oven in the kitchen. If this experiment is performed in a very cold region, it is suggested to have a heater or near the general heater of the house to maintain high temperature above room temperature. Anaerobic conditions favor ethanol production over other metabolic pathways.

It is important that the pH, temperature, and time are varied to investigate their effects on the percentage of ethanol produced (Table 2.2).

2.2.4.1 Sampling and Replicates

Samples are taken twice a day for 72 hours. These samples are immediately filtered using 0.22 µm Nylon membranes and stored in refrigeration for subsequent analysis. Frequent sampling allows for detailed monitoring of the fermentation process.

2.3 Analytical Techniques

2.3.1 Glucose Quantification at Home Using Hydrometer Method

Hydrometer measures the specific gravity of a liquid compared to water. Ethanol is less dense than water so as fermentation converts glucose to ethanol, the specific gravity drops. So, to achieve this measurement, you need a hydrometer (you can buy from Amazon or brewing supply stores), a graduated cylinder or test jar, and your fermented liquid. Figure 2.2 shows a typical home safe hydrometer.

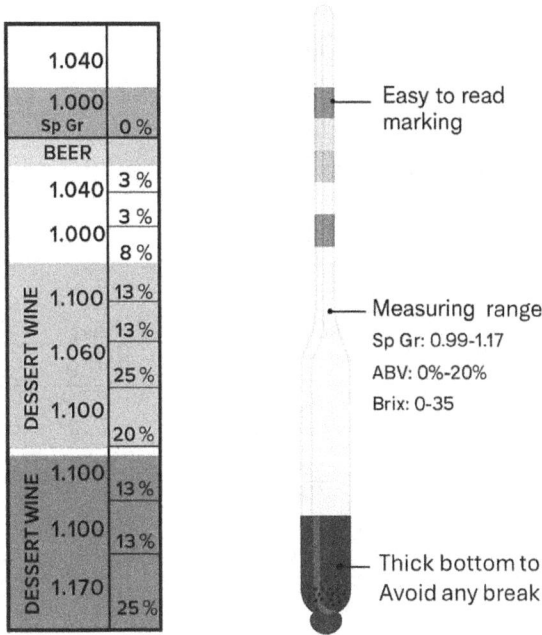

FIGURE 2.2
Hydrometer for glucose quantification.

2.3.2 Procedure

Before fermentation, take the original specific gravity (density, ρ_0) of the liquid (e.g., 1.050). Then after fermentation, take the final gravity reading (density, ρ_f). Then calculate alcohol by volume as given in Equation 2.1.

$$\text{Eth}_v\,(\%) = \left(\rho_0 - \rho_f\right) \times 131.25 \qquad (2.1)$$

Advantages: It is simple, affordable, and accurate for most homebrewing.

Limitations: Only works for fermented beverages (not distilled spirits), and it assumes all gravity change is from ethanol (other factors like residual sugars can skew it slightly).

Hence, to purify the ethanol, then it is necessary to perform distillation by boiling the fermented liquid in pot at 78°C (use simple thermometer that can be bought at online store) and collecting the vapor. Then allow it to condense to pure ethanol.

2.4 Discussion and Student Tasks

The results should be discussed in the context of the fermentation conditions and the metabolic activity of *Saccharomyces cerevisiae*. Factors such as temperature, pH, and nutrient availability can significantly impact ethanol production.

2.4.1 Questions

What would be the possible causes of low ethanol yield?

_____

_____

_____

_____

_____

_____

What could be the possible problems of slow fermentation, and how can they be avoided?

_____

_____

_____

_____

_____

_____

It is possible to have inconsistent results during this experiment; can the students describe what could cause variability of the results?

_____

_____

_____

_____

_____

_____

2.5 Safety Precautions

2.5.1 General Lab Safety

- Wear appropriate personal protective equipment (PPE), including gloves and eye protection.
- Handle chemicals with care and follow safety guidelines.
- Dispose of waste materials properly.

2.5.2 Specific Precautions

- Ensure proper ventilation during fermentation to avoid the buildup of CO_2.
- Use caution when handling acids and bases for pH adjustment.
- Be aware of the flammability of ethanol and avoid open flames.

2.6 Conclusion

The experiment details the protocol for conducting ethanol production experiments at home. By following these guidelines, one can gain a practical understanding of the alcoholic fermentation process and the factors that influence ethanol yield and efficiency. Further research and optimization can be performed to improve the process and explore different substrates and yeast strains.

References

1. Y. Lin and S. Tanaka, "Ethanol fermentation from biomass resources: Current state and prospects," *Applied Microbiology and Biotechnology*, vol. 69, no. 6, pp. 627–642, Feb. 2006, https://doi.org/10.1007/S00253-005-0229-X.
2. R. Morales-Rodriguez, K.V. Gernaey, A.S. Meyer and G. Sin, "A mathematical model for simultaneous saccharification and co-fermentation (SSCF) of C6 and C5 sugars," *Chinese Journal of Chemical Engineering*, vol. 19, no. 2, pp. 185–191, Apr. 2011, https://doi.org/10.1016/S1004-9541(11)60152-3.
3. M. Jayakumar, S. Kuppusamy Vaithilingam, N. Karmegam, K.B. Gebeyehu, M.S. Boobalan and B. Gurunathan, "Fermentation technology for ethanol production: Current trends and challenges," *Biofuels and Bioenergy: A Techno-Economic Approach*, pp. 105–131, Jan. 2022, https://doi.org/10.1016/B978-0-323-90040-9.00015-1.

3

3D Design of Continuous Millireactor and Monoliths for Bioenergy Application

Summary

Teaching and learning of 3D structures applied to reaction engineering has become essential to illustrate the transition from lab-scale practices to industrial applications. This practical manual proposes to enhance the students' education by integrating the design and fabrication of 3D structures that can function as catalysts and millireactors for application in reaction engineering. Autodesk Inventor Pro was used to model 3D structures and render them for 3D printing in 3D printers, and filaments are easily available in online stores. Diverse monolithic shapes such as hexagonal, triangular, rectangular, and circular were explored. These printed catalyst structures can be used to propagate chemical reactions. The millireactor is a serpentine reactor which allows for continuous reaction. The students are encouraged to attempt all the tasks to improve understanding.

3.1 Practice 1: Installing and Configuring Solid-Works

3.1.1 Objectives

- Knowing the system and software requirements to run Inventor® correctly.
- Installing Autodesk® product components is necessary for the following practices.
- Configuring Inventor® options and preferences to customize the working interface to make it friendly for the user.

3.1.2 Background

Inventor® is computer-aided design software that allows you to create 3D models of mechanical parts and assemblies, as well as extract drawings and

DOI: 10.1201/9781003374503-3

information for manufacturing. 3D design in Inventor® is based on parametric operations and automatic or user-defined relationships, which capture the design intent and facilitate its modification.

3D design in Inventor® has various applications and advantages, such as virtual prototyping, performance optimization, cost and time reduction, and integration with other engineering and manufacturing systems.

To perform this practice, it is necessary to install and configure Inventor® correctly on the computer, following the minimum system requirements, license type, and installation steps. It is also convenient to customize the configuration options according to the user's preferences.

3.1.3 How to Install Autodesk Inventor Professional in Your Computer

1. Go to the "www.autodesk.com" website.
2. Select Products and **Inventor Pro**. You can have 30 days free trial.
3. Within the Autodesk® website, click on the "Start button".
4. Click "download free trial". It will show you another prompt, asking what you will use the software for.
5. Click "🎓 Education" and then you have to sign in to the education community.
6. Click **Inventor Pro** and you can see the options that best suits your condition. If your school is not authorized, you can still have free access for 30 days.
7. Fill the form online to validate that you are a student. The verification process can take 48 hours and if you have educational access to Autodesk products, you must wait one year from the last submission before you can make a new request.
8. After verification, you will have all the software available. Search for Inventor Professional® and click on "Get Product", and it will start the download. Follow the instructions to finish the installation.

3.2 Practice 2: Design of 3D Models of Monoliths of Different Shapes Using Inventor®

3.2.1 Objectives

At the end of the practice using this manual, the student will be able to:

- Understand the basic principles of 3D modeling in Inventor®.

- Select the appropriate materials and dimensions for the monoliths.
- Design the monolith in 2D and extrude the monolith in 3D.

3.2.2 Quiz

1. What are 3D-printed monoliths, and why are they preferred as high surface area structures for catalyst supports, functional catalysts, or bioreactor scaffolds?

   _____

   _____

   _____

   _____

   _____

2. Discuss the various advantages of 3D printed monoliths compared to pellets of different shapes for distillation column application.

   _____

   _____

   _____

   _____

   _____

3.2.3 Theoretical Framework

The advent of 3D printing enables the fabrication of catalyst supports and catalytic structures with precisely engineered geometries and optimized dimensions. Unlike conventional catalysts, typically shaped as densified pellets to improve mechanical strength in packed-bed reactor configurations, 3D-printed monoliths offer a distinct structural and functional advantage. Their interconnected channels and high geometric surface area significantly enhance mass and heat transfer, facilitating faster reaction kinetics and higher conversion efficiencies.

Monolithic architectures promote uniform flow distribution and minimize pressure drop, addressing common limitations in conventional packed beds such as channeling and high energy input for fluid movement. These structures ensure consistent contact between reactants and the catalytic surface, which is critical for performance in continuous flow processes. Furthermore, the ability to tailor the internal geometry at the microscale enhances the accessibility of active sites and improves spatial control over reaction zones.

3D-printed monoliths often demonstrate superior thermal and mechanical stability under harsh operating conditions, making them suitable for high-temperature and high-pressure applications. Their resistance to attrition and deactivation extends catalyst life and improves overall process reliability. The improved efficiency in mass and heat transfer can also lead to significant energy savings and reduced operational costs. In addition to chemical synthesis, 3D-printed monolithic catalysts are increasingly applied in environmental remediation, such as the removal of greenhouse gases and other pollutants, due to their tunable porosity and functional surface modification potential.

3.2.4 Circular Geometry

1. Open a new project in Inventor® and select "Part (.ipt)," as shown in Figure 3.1.
2. Go to "3D Model" tab and select "New 2D sketch" to start drawing the basic geometry of the objective structure and design of the filter. Start it choosing a plane that you will use to draw, in this case it will be selected XY Plane (Figure 3.2).
3. Once you have chosen the plane, start designing the first sketch. For this example, select "Circle" and trace it starting from the center of the plane (origin), in order to facilitate the design. The diameter of this circumference will be 30 mm (Figure 3.3).
4. Again, on "3D Model" tab select "Extrude." Click on the surface of the circumference and proceed to extrude it, which should go to

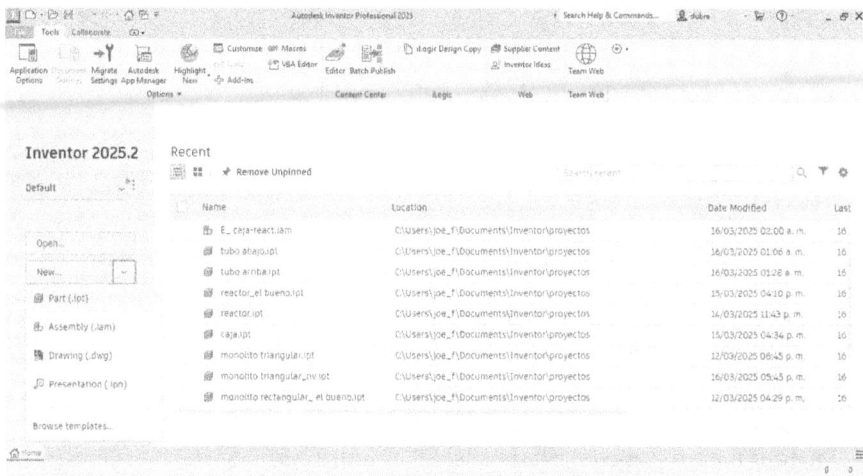

FIGURE 3.1
Initial menu shown in Inventor®.

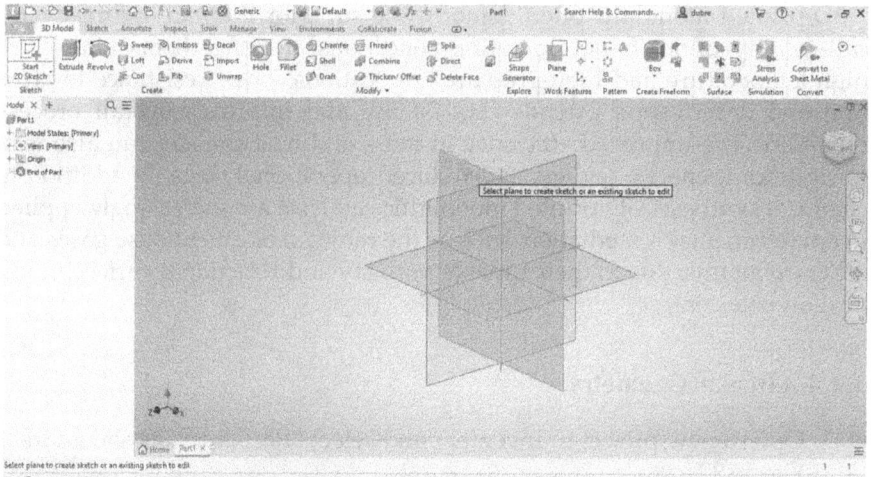

FIGURE 3.2
Selection of the work plane.

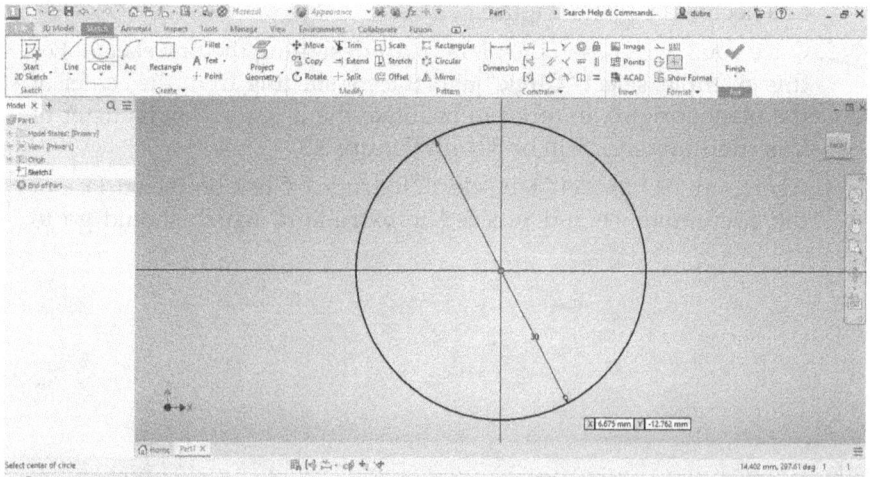

FIGURE 3.3
Circumference as a base of the design.

DEFAULT direction, and it must measure 20 mm. (Please check more details of the extrude in Figure 3.4.)

5. Now, to continue working on the original sketch and add details to the monolith (the holes on it); right click on "Sketch 1" showed on the right menu. Choose "Share sketch" (Figure 3.5).

6. To obtain filters with a circular shape, draw a new line of 14 mm starting from the origin of the initial circular shape in any direction.

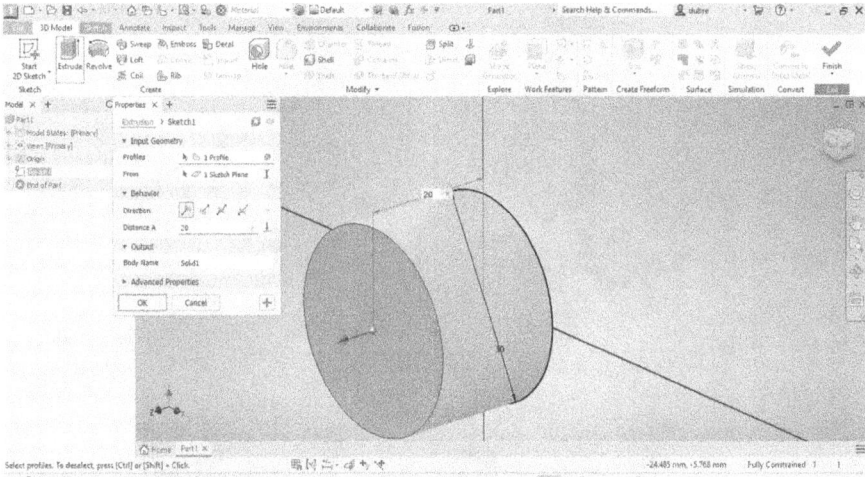

FIGURE 3.4
Circular extrusion.

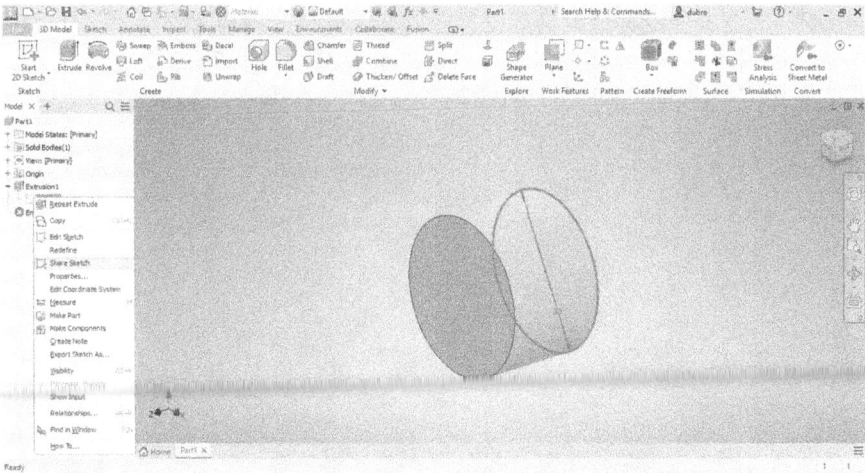

FIGURE 3.5
Sharing the initial sketch.

Trace a circle of 1.5 mm diameter at the rear end of the line. Repeat this step using guidelines at the following dimensions: 11, 8, 5, and 2 mm as shown in Figure 3.6. Avoid making these strokes in the same direction.

7. Finish "Sketch 1" and return to the "3D Model" tab to extrude the newest small circles. In this case, use the DEFAULT direction; it must measure 20 mm and select CUT. Repeat this step with all the small circles drawn on the sixth instruction (Figure 3.7).

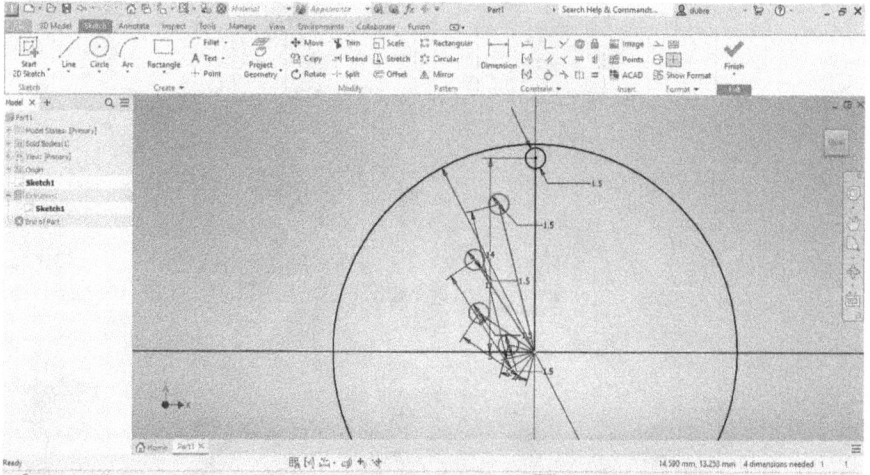

FIGURE 3.6
Base internal geometry of the monolith.

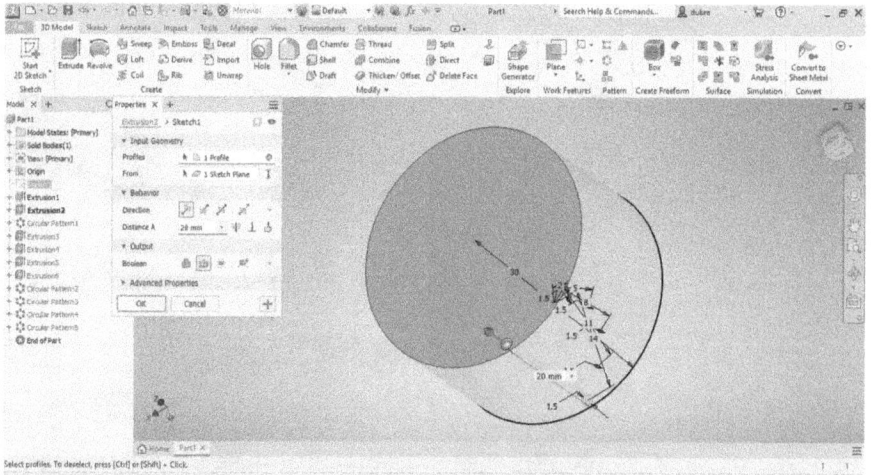

FIGURE 3.7
Cut of the selected geometry for the monolith.

8. Selecting one of these small circular extrusions, choose "Circular pattern" in order to be able to reproduce this cut along the first circumference, as shown in Figure 3.8. In this case, consider repeating this geometry 30 times; however, it is important to always consider the diameter of the filter and the separation between the holes on it.

9. Please repeat the previous steps, in order to obtain the expected structure, creating the following pattern (Figure 3.9).

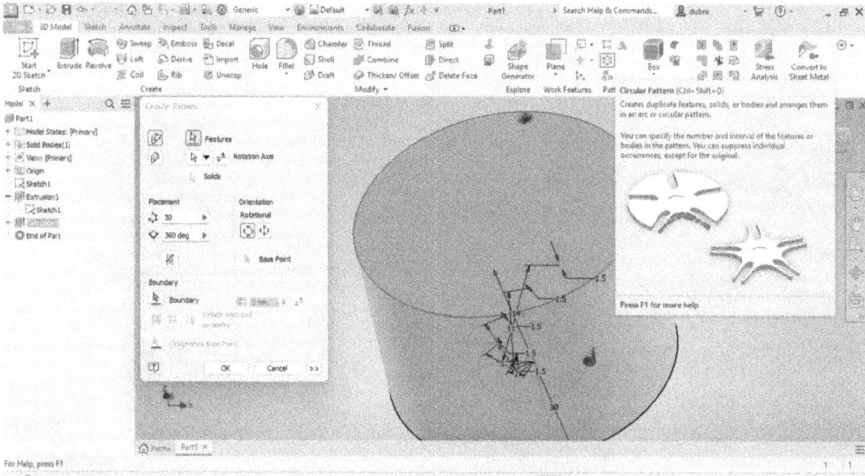

FIGURE 3.8
Circular pattern on filter.

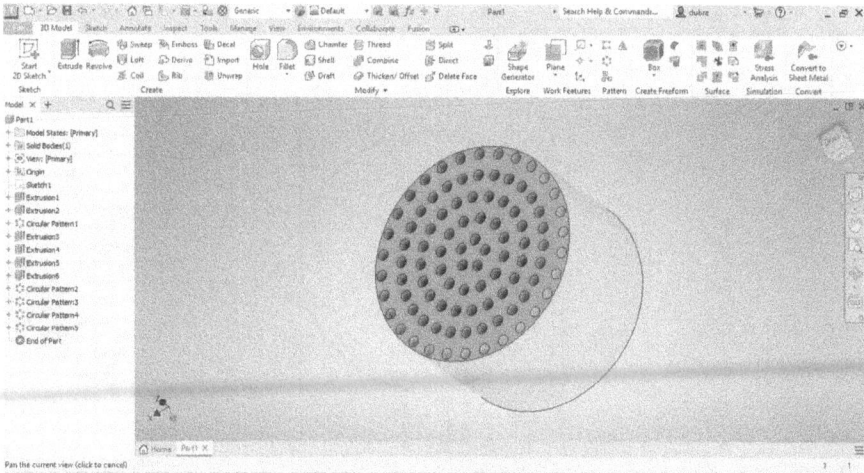

FIGURE 3.9
Finished circular pattern monolith.

3.2.5 Square Geometry

1. Repeat the first four steps of the previous geometry. Add to this initial sketch an internal circumference which has to measure 29.5 mm. Additionally, select "Two point center" option (Figure 3.10).

2. Starting from the center of the plane, draw a square which has to measure 30 mm each side. The result is shown in Figure 3.11.

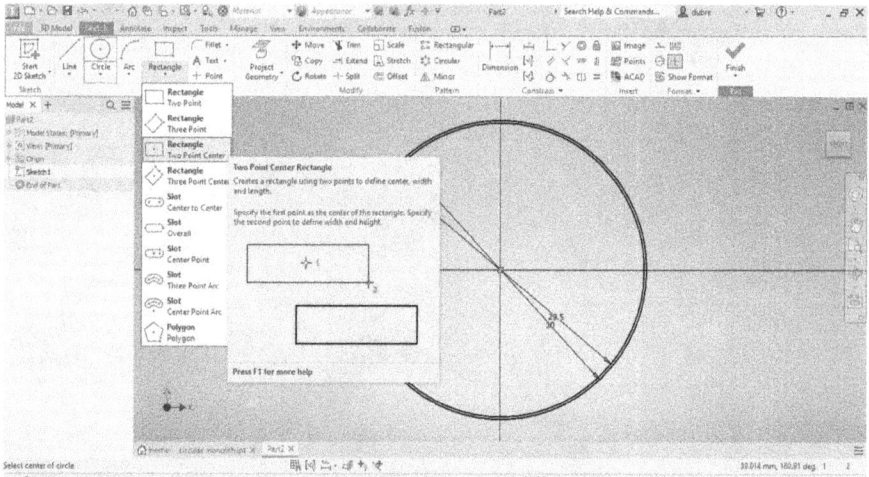

FIGURE 3.10
Circumferences as a base of the monolith.

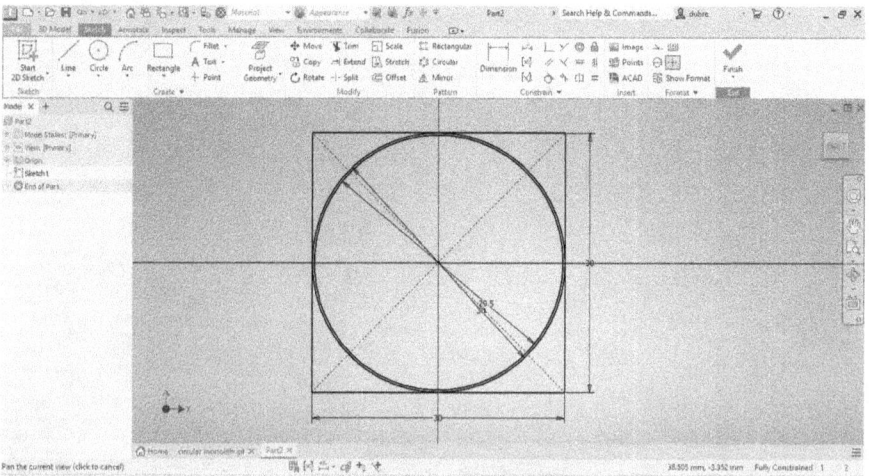

FIGURE 3.11
Essential base geometry.

3. As a next step, it will be necessary to build a mesh in the initial sketch that will originate the square geometry that is sought for this monolith. Selecting the "Two-point rectangle" option, trace a square starting from one of the edges of the bigger square drawn on the previous step; it has to measure 1.5 mm. Please see Figure 3.12.

4. To create the mesh based on the squares drawn in step 3, choose "Rectangular pattern" in order to be able to reproduce it along the

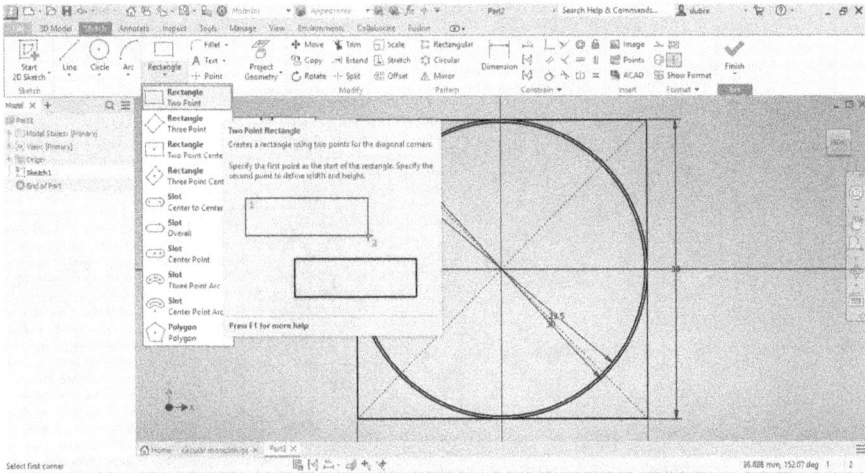

FIGURE 3.12
Initial sketch to star the mesh.

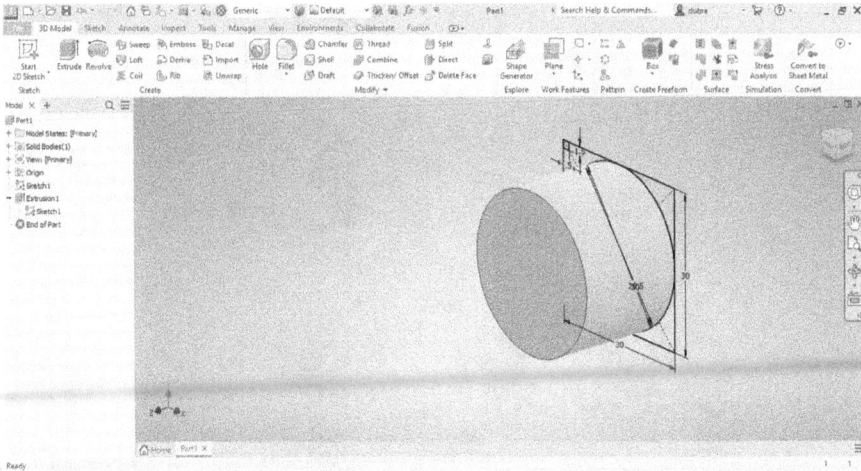

FIGURE 3.13
Meshing of 2 mm square.

first square. See Figures 3.13 and 3.14 to know more details about this instruction for each small square.

5. Click again on Extrude option. Select each square surface created by the mesh (step 4) that proceeds to extrude it, which should go to DEFAULT direction, and it must be 20 mm and select CUT. Repeat this step with all the small squares drawn. Please check more details of the extrude in Figure 3.15.

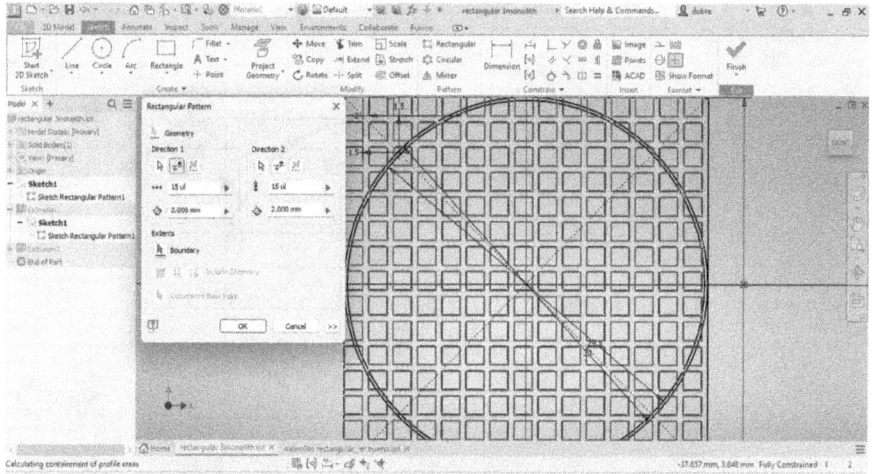

FIGURE 3.14
Meshing of 1.5 mm square.

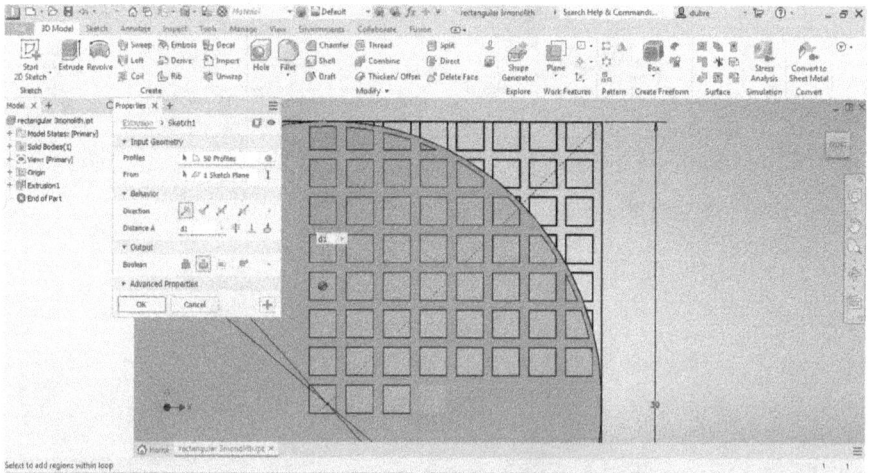

FIGURE 3.15
Final mesh pattern.

6. Repeat the previous steps, in order to obtain the expected structure, creating the following pattern (Figure 3.16).

3.2.6 Triangle Geometry

1. Repeat the first two steps of the previous geometry. This time, extend the square until it reaches 32 mm (Figure 3.17).

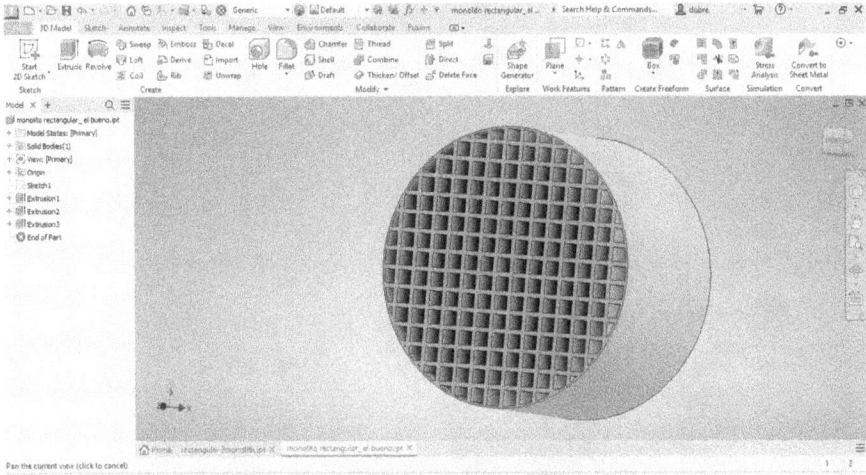

FIGURE 3.16
Finished square pattern monolith.

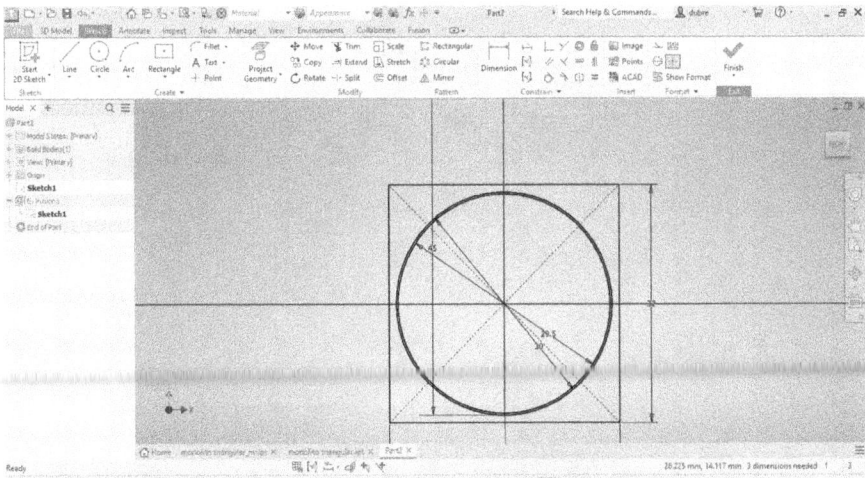

FIGURE 3.17
Concentric circles and the formation of monolith geometry.

2. Again, the objective is to create a triangle mesh based on the rectangle drawn in the previous step. To do that, create a triangle starting from one of the edges of the bigger square and choosing "Line," the measurements will be 1 mm and 60°. Now, leave a 0.25 mm space and trace an inverter triangle (go to Figure 3.18).

3. The purpose of these triangles is to prepare a guideline to select "Rectangular pattern." This option will be able to reproduce the

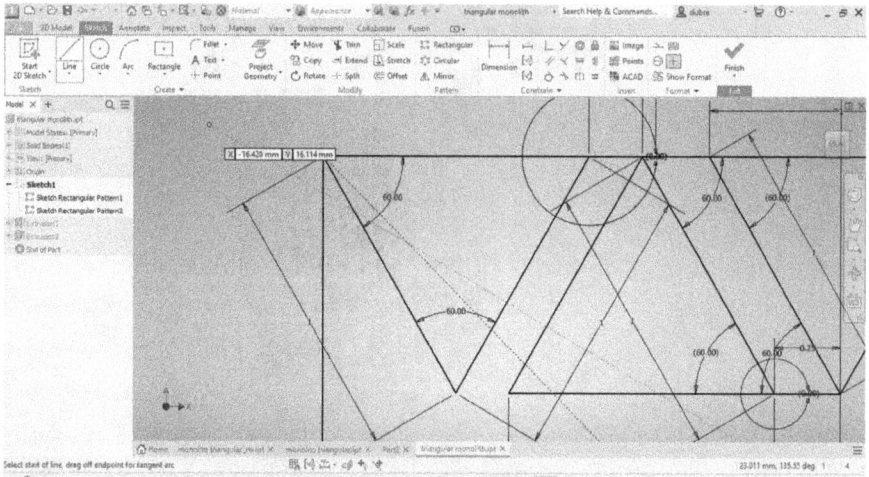

FIGURE 3.18
Initial mesh with triangular geometry.

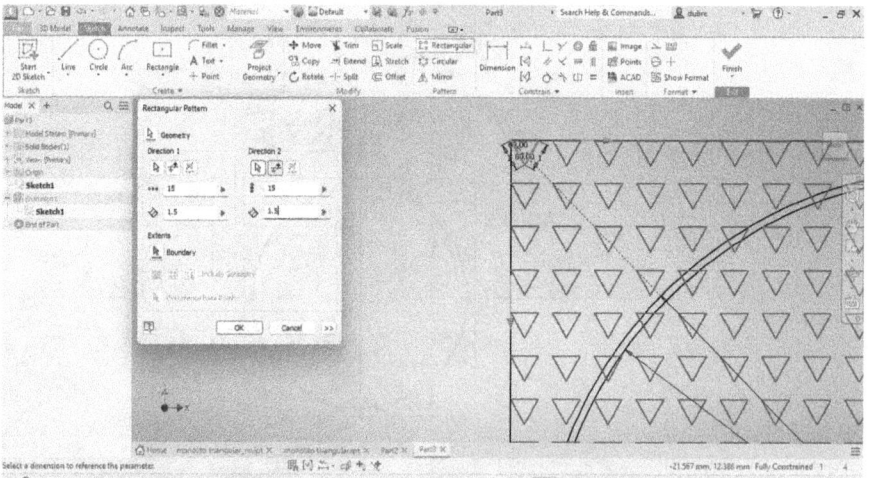

FIGURE 3.19
Meshing the monolith.

figures drawn along the square. See Figure 3.19 to know more details about the last instruction.

4. Repeat step 3 for the inverter triangle until obtain a meshing like Figure 3.20.

5. Finally, the sketch shown in Figure 3.21 will be obtained.

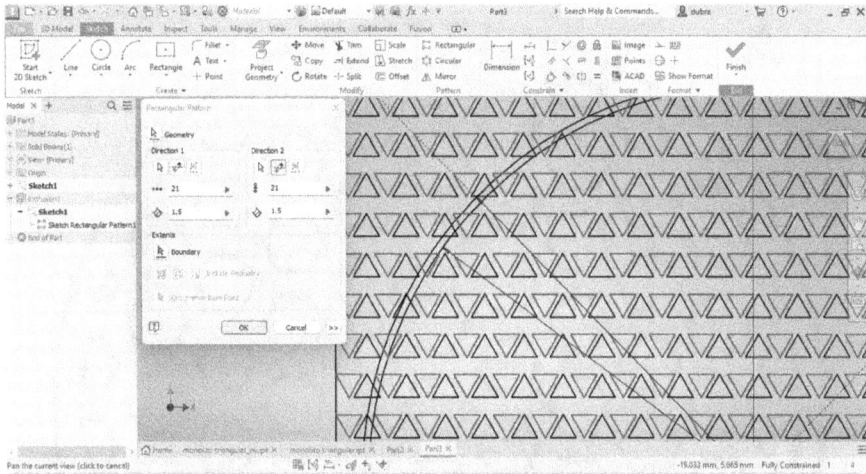

FIGURE 3.20
Initial mesh with triangular geometry.

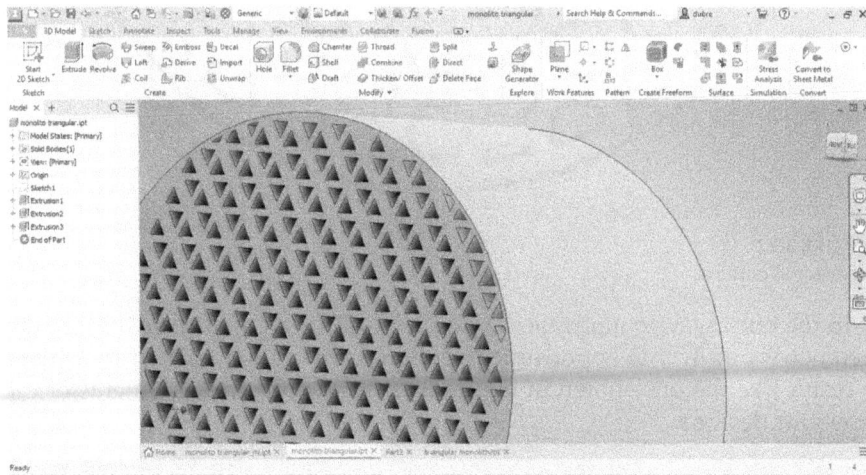

FIGURE 3.21
Finished triangular pattern monolith.

3.2.7 Student Challenge

The students should try to design a monolith with hexagonal geometry (Figure 3.22). Present it in 3D in Inventor® and describe the step by step methods to realize the drawing. Use similar dimensions as those used in the previous sketches.

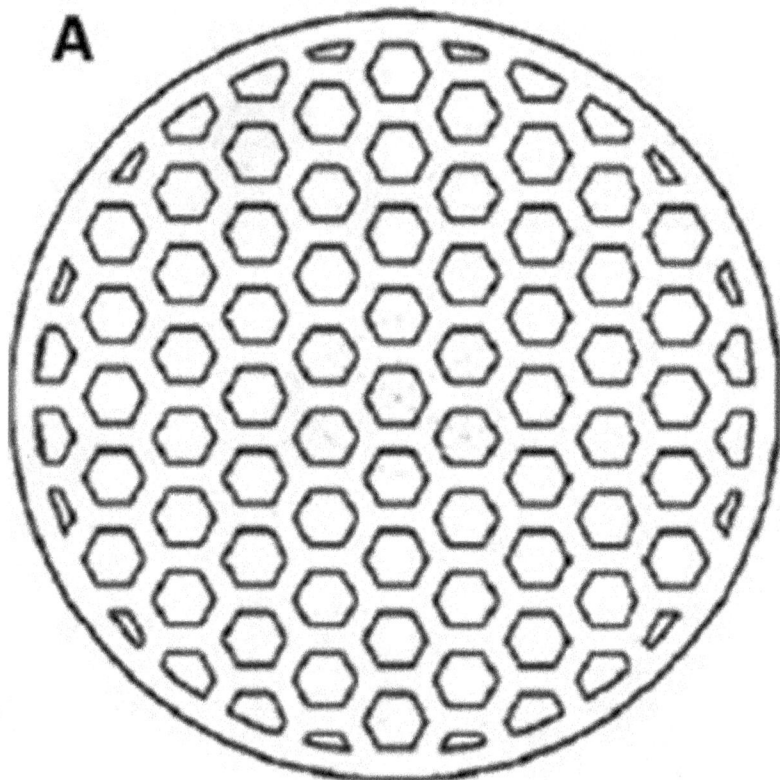

FIGURE 3.22
Hexagonal pattern.

In the same way, to motivate the practice of 3D design, recreate the circular geometry but in this opportunity applying rectangular pattern to achieve a result like the one shown in Figure 3.23. Follow the measurements of the previous designs.

3.3 Practice 3: Design of 3D Serpentine Millireactor in Inventor®

3.3.1 Objectives

At the end of Practice 3, the student will be able to:

- Design millireactors and understand the concept in continuous flow reaction systems.
- Select the appropriate materials and dimensions for 3D millireactor design.

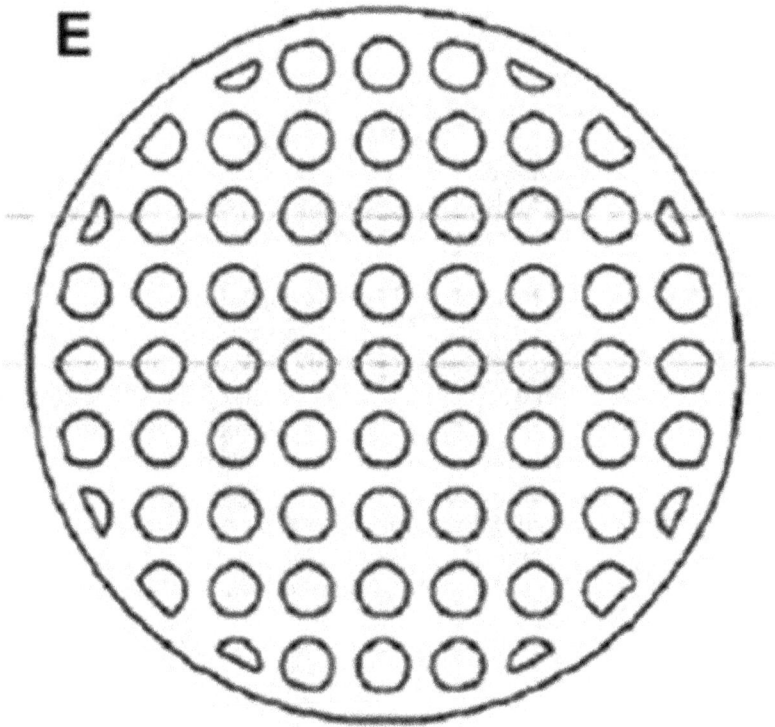

FIGURE 3.23
Circular geometry using rectangular pattern.

3.3.2 Theoretical Framework

Reactions involving homogeneous catalysts, such as transesterification, esterification, and other liquid-phase transformations, are often constrained by phase equilibrium, catalyst recovery challenges, and thermal sensitivity. These limitations can be addressed effectively through the design and implementation of continuous flow reactors optimized for low-temperature operation.

One common example is the production of biodiesel via transesterification, where triglycerides from various feedstocks (e.g., vegetable oils, animal fats, waste cooking oils) are reacted with alcohols in the presence of a homogeneous base catalyst, typically sodium or potassium hydroxide. Industrial processes prefer continuous operation due to its efficiency, scalability, and tighter process control. Conversion efficiencies exceeding 95% have been routinely reported using alkaline homogeneous catalysts under mild temperature conditions.

The choice of reactor type and materials must align with both the chemical requirements and the mechanical demands of the process. For low-temperature homogeneous catalytic reactions, the reactor must ensure effective mixing, adequate residence time, and minimal thermal loss, while maintaining chemical compatibility with the reaction mixture.

Traditionally, continuous stirred tank reactors (CSTRs), fixed-bed reactors (modified for homogeneous phase operation), and fluidized beds have been adapted for such systems. However, with advances in additive manufacturing, particularly fused deposition modeling (FDM), polymer-based millireactors are increasingly being explored. These reactors offer low-cost, customizable, and modular platforms suitable for academic and pilot-scale research.

In this context, materials like Acrylonitrile Butadiene Styrene (ABS) and Polylactide (PLA) have been used to fabricate small-scale continuous reactors. A serpentine millireactor design was adopted in this project to promote mixing and residence time across more than 1 meter flow path. The compact geometry enables plug flow behavior and better control of reaction parameters such as temperature and mixing. Moreover, students can fabricate and stack multiple reactor modules to extend path length, enhancing conversion and throughput while maintaining manageable pressure drops.

These 3D-printed systems offer a flexible platform for studying continuous homogeneous catalysis, particularly when conventional reactor materials or designs are cost-prohibitive or unsuitable for experimental optimization. While these polymer-based reactors may not be suitable for high-temperature or corrosive environments, they serve as effective tools for low-temperature continuous processes where rapid prototyping and adaptability are critical.

3.3.3 3D Design of Serpentine Millireactor

Due to the level of complexity, the different components of the millireactor will be designed separately. Once completed, all the components will be assembled together.

3.3.3.1 *Serpentine Millireactor Casing or Base*

1. To start, repeat the first two steps of the previous circle monolith geometry. Sketch a rectangle starting from the origin of the plane by selecting "Two point center" option. It should measure 190 mm of length and 48 mm of height as shown in Figure 3.24.

2. Trace a line starting from the lower left corner and extend it along the pathway marked by the initial rectangle. Mark the distances following this order: 1.5, 1, 1.5, and 1.5 mm, see Figure 3.25. Repeat it on the upper right corner, these lines will be used as references in the next step.

3. Finish "Sketch 1" and return to the "3D Model" tab to extrude the rectangle. In this case, use the DEFAULT direction; it must measure

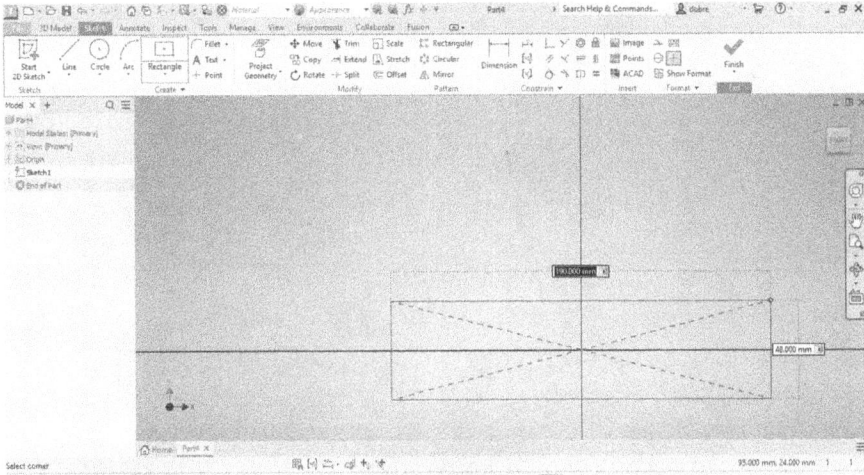

FIGURE 3.24
Initial sketch.

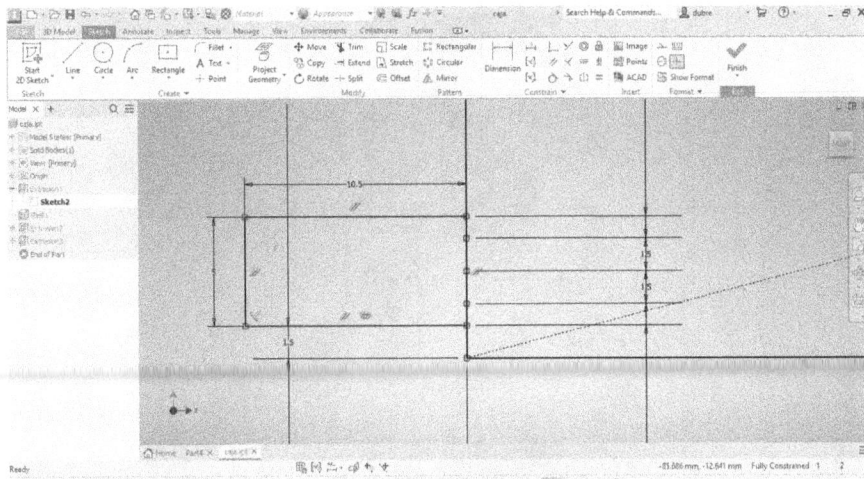

FIGURE 3.25
Establish guidelines.

5 mm (Figure 3.26). In the same tab, select "Shell" (1 mm to the center) and remove the front face. The result should look like Figure 3.27.

4. Start a new sketch on the right side of the new case, starting from the guidelines traced in the second step. Draw a square of 5 mm on each side, followed by two diagonal lines to find the center of it. The center of the square will be the point of reference to define a new circle

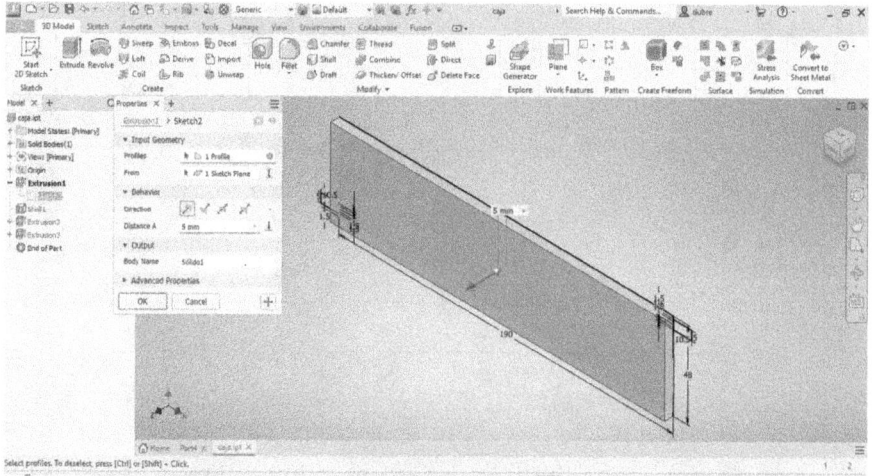

FIGURE 3.26
Case extrusion.

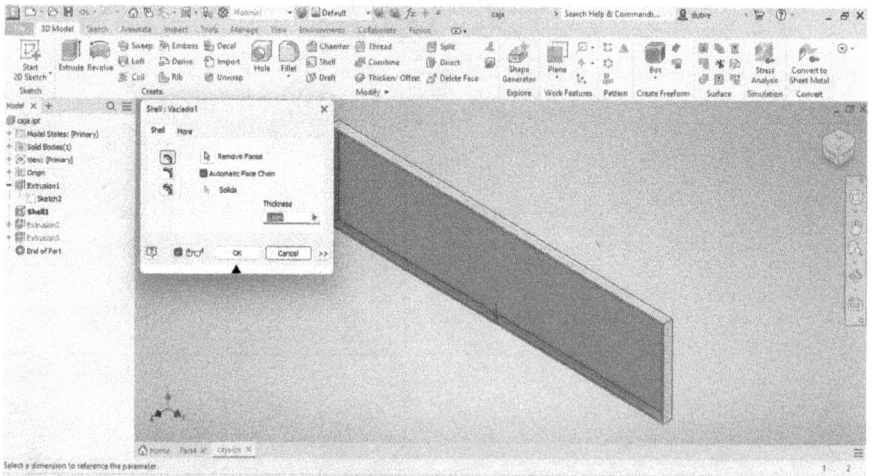

FIGURE 3.27
Case design.

    of 3 mm in diameter, use Figure 3.28 as a reference. Repeat this step on the left side of the structure.

5. Go to the "3D Model" tab to extrude the newest small circles. Apply REVERSE direction; it must measure 1 mm and select CUT (Figure 3.29).

6. The resulting structure should look like Figure 3.30.

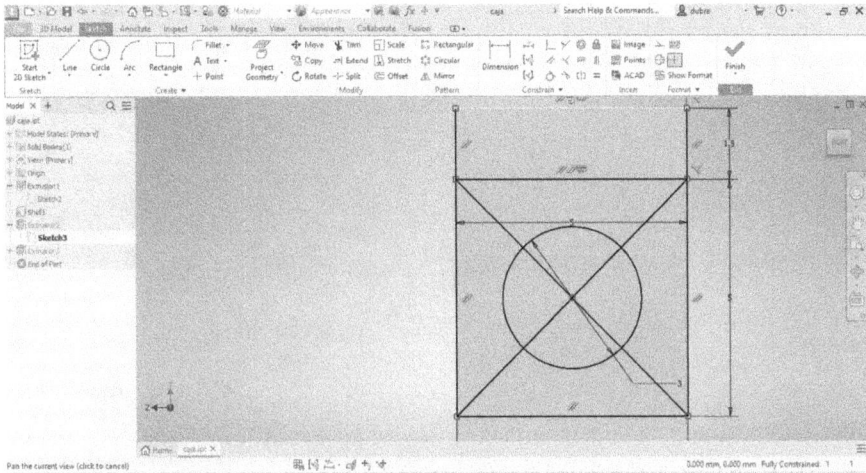

FIGURE 3.28
Serpentine inlet and outlet.

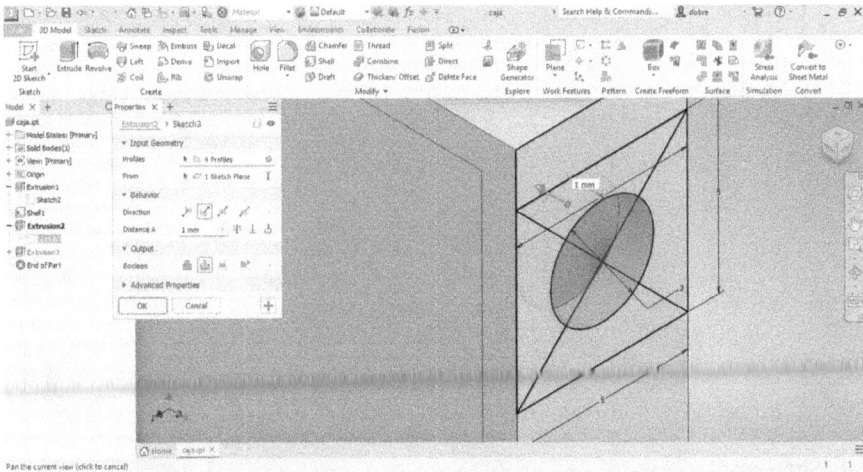

FIGURE 3.29
Cut of the lateral surface of the bioreactor.

3.3.3.2 *Serpentine Channel*

7. Repeat again the first two steps of the previous section. Trace a new line (4 mm) starting from the upper right corner in the direction marked by the constructed rectangle. From this point, trace a perpendicular line (180 mm) addressed to the opposite side of the rectangle. Encore it again, from one side to the other, until you finally obtain the initial structure of a serpentine. Please see Figure 3.31.

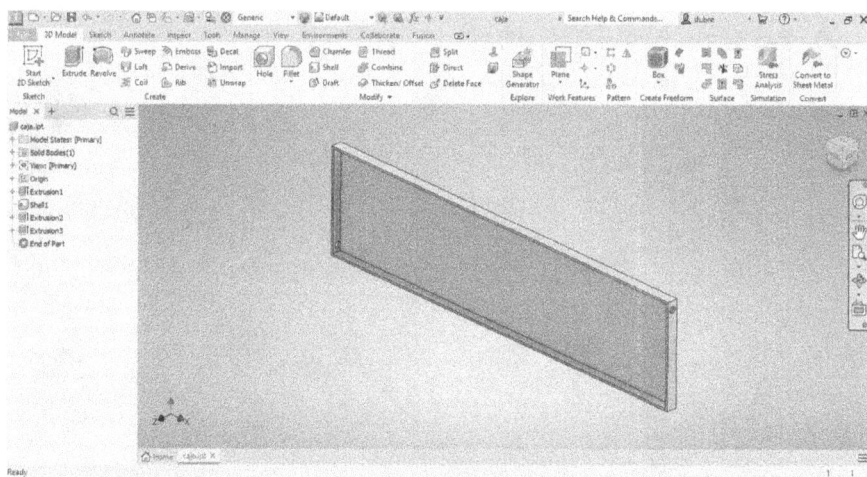

FIGURE 3.30
Finished structure.

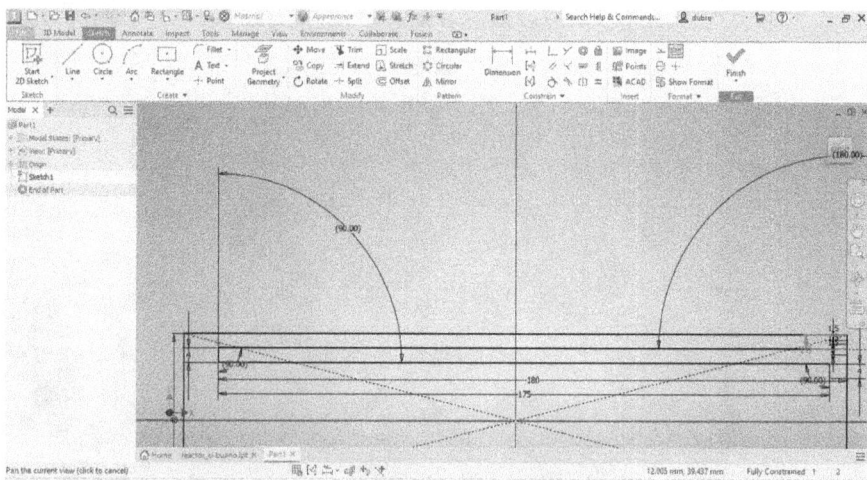

FIGURE 3.31
Initial sketch of serpentine.

8. On "Sketch" tab, look for the "Fillet" the option and apply it to all the edges generate on the serpentine. It has to measure 2 mm (Figure 3.32). The final sketch has to look like Figure 3.33.

9. Click on the "3D Model" tab and go to the "Work Features" panel and click on "Normal to Curve at Point," to create a new plane centering on the 4 mm line traced on step 7 and perpendicular to the XY plane. Repeat it for the opposite side (Figure 3.34).

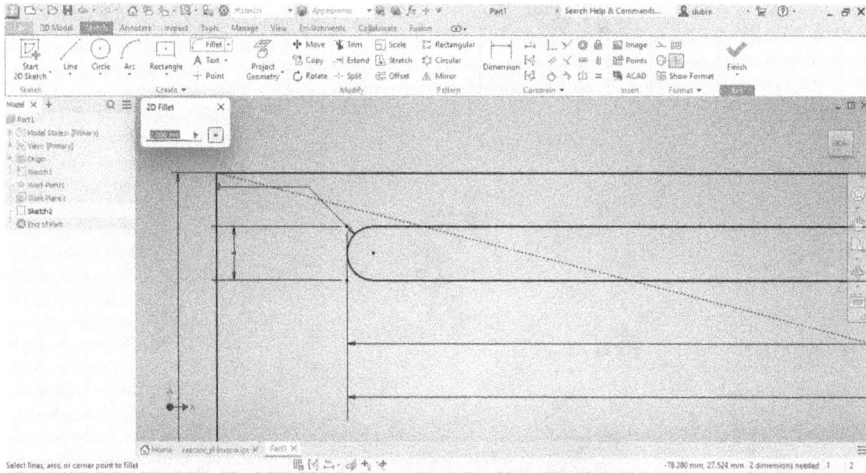

FIGURE 3.32
Joining the edges of the serpentine.

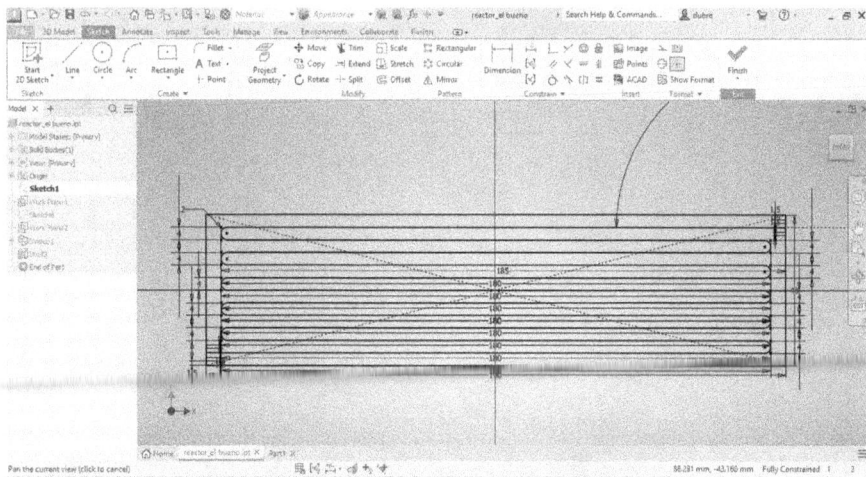

FIGURE 3.33
Final sketch of serpentine.

10. Draw a circle with a diameter of 3 mm, starting from the center of the newly created planes Figure 3.35.

11. Again, go to the "3D Model" tab and choose the "Sweep" option. To complete the instruction, it will be required to select a profile (the circle drawn in the previous step) and a path. See Figure 3.36 to know more details about the last instruction.

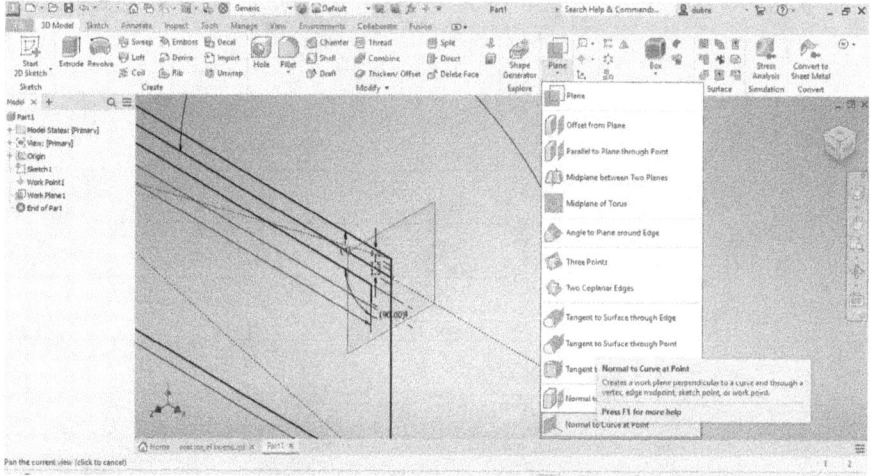

FIGURE 3.34
Creation of perpendicular planes to design the serpentine inlet and outlet.

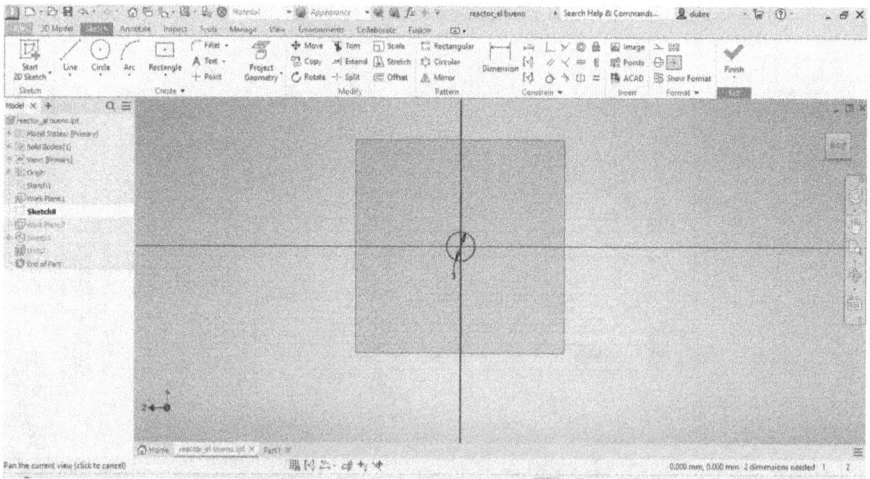

FIGURE 3.35
Initial sketch of serpentine inlet and outlet.

12. In the same tab, select "Shell" (0.2 mm to the diameter of the circumference) and remove the front face of both circular faces. As a result, it will obtain a pipe that should look like Figure 3.37.

13. The resulting structure should look like Figure 3.38.

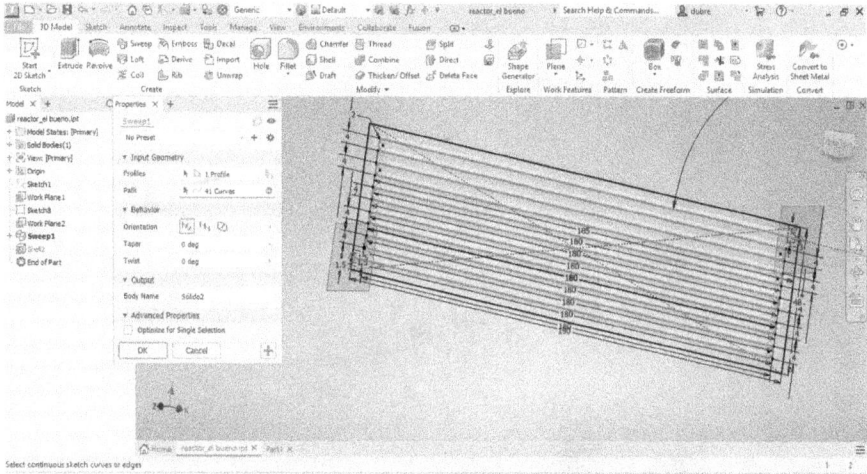

FIGURE 3.36
Pipeline layout.

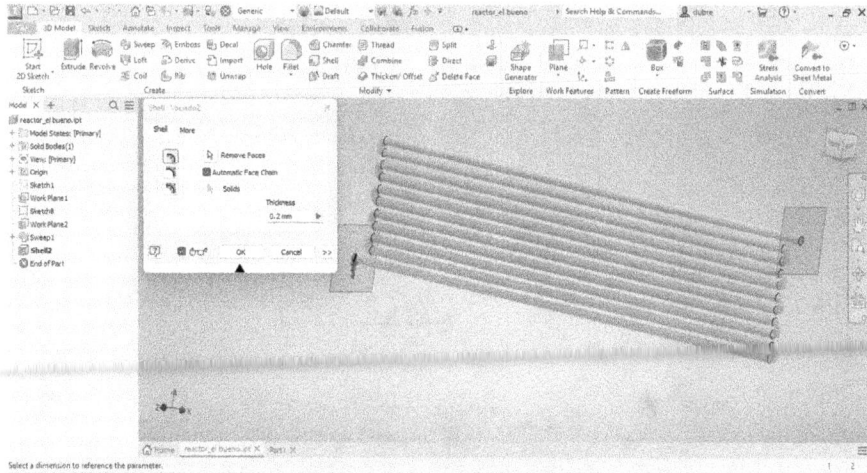

FIGURE 3.37
Internal pipeline design.

3.3.3.3 Inlet Pipe

14. Open a new project in Inventor® and select "Piece (.ipt)." Go to "3D Model" tab and click on "New 2D sketch" to start drawing the basic geometry of the structure. This time, choose a plane ZY Plane and trace the sketch shown in Figure 3.39.

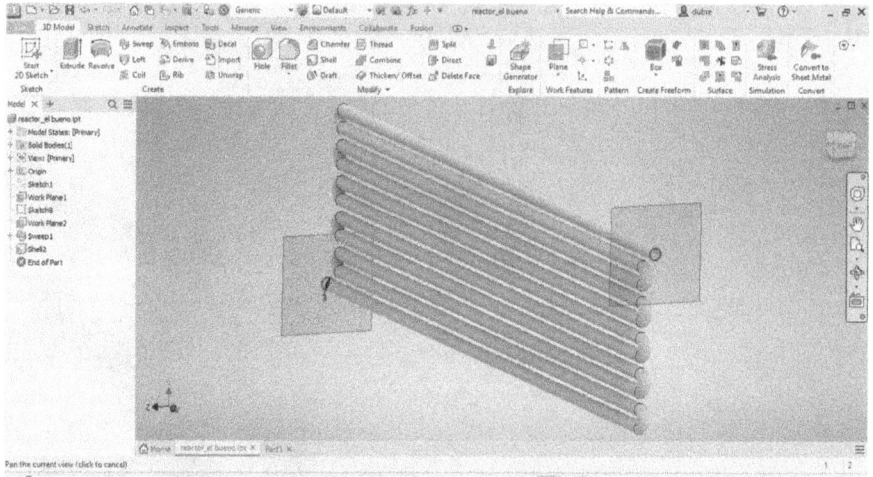

FIGURE 3.38
Finished structure.

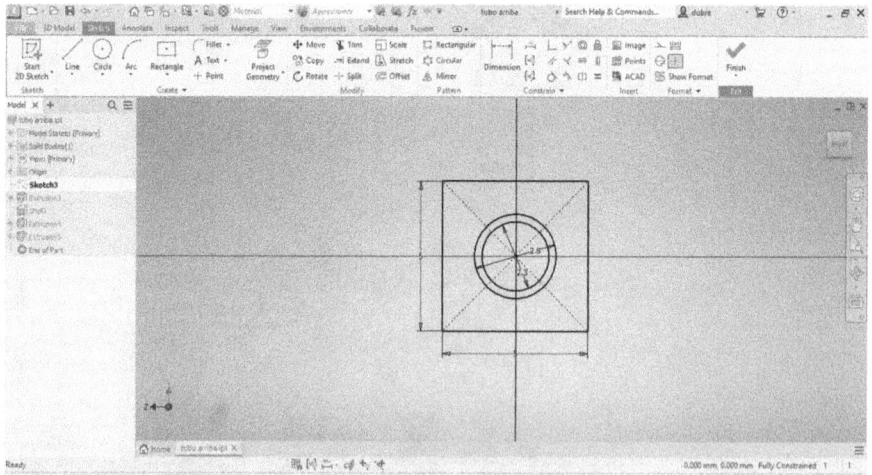

FIGURE 3.39
Initial sketch of the inlet pipe.

15. Finish "Sketch 1" and go to the "3D Model" tab to extrude the square. In this case, use the DEFAULT direction; it must measure 10.5 mm. In the same tab, select "Shell" (1 mm to the center) and remove the front face. The result should look like Figure 3.40.

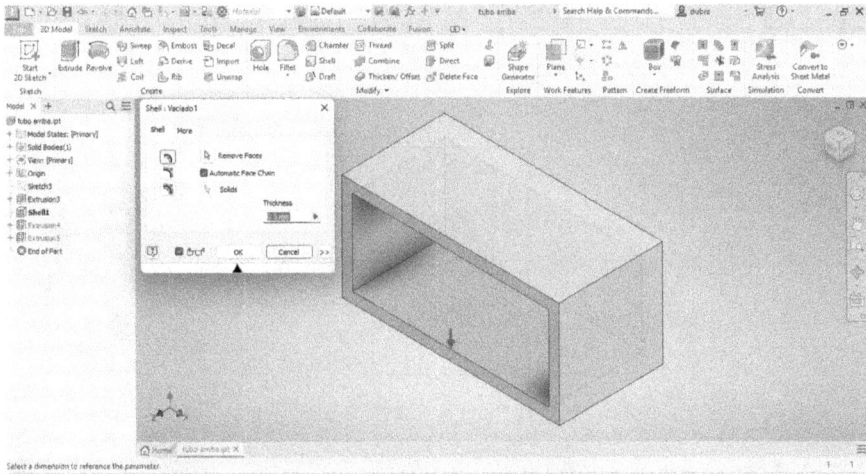

FIGURE 3.40
Extrusion and shelling of the external structure.

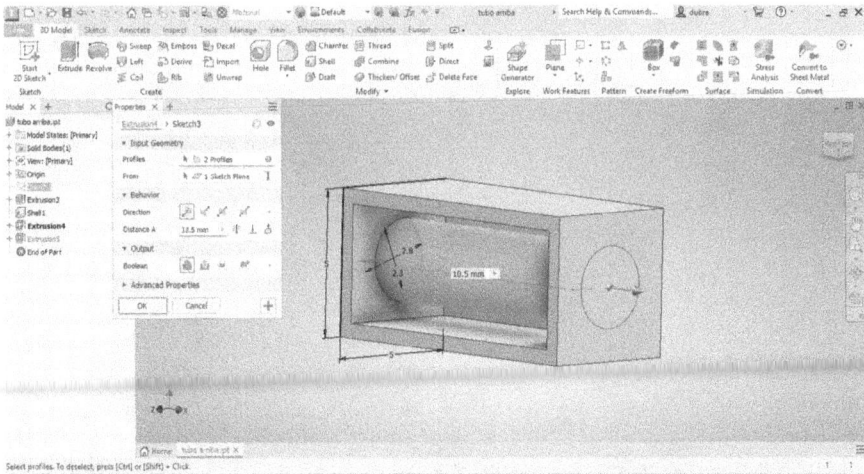

FIGURE 3.41
Extrusion of the external face of the pipe.

16. Right click on "Sketch 1" showed on the right menu and choose "Share sketch," to continue editing the initial sketch. Now, extrude the 3-mm diameter circle, using the DEFAULT direction; it must measure 10.5 mm (Figure 3.41).

17. Extrude the 2.8-mm diameter circle, using the DEFAULT direction. It must measure 10.5 mm and select CUT (Figure 3.42).

18. The resulting structure should look like Figure 3.43.

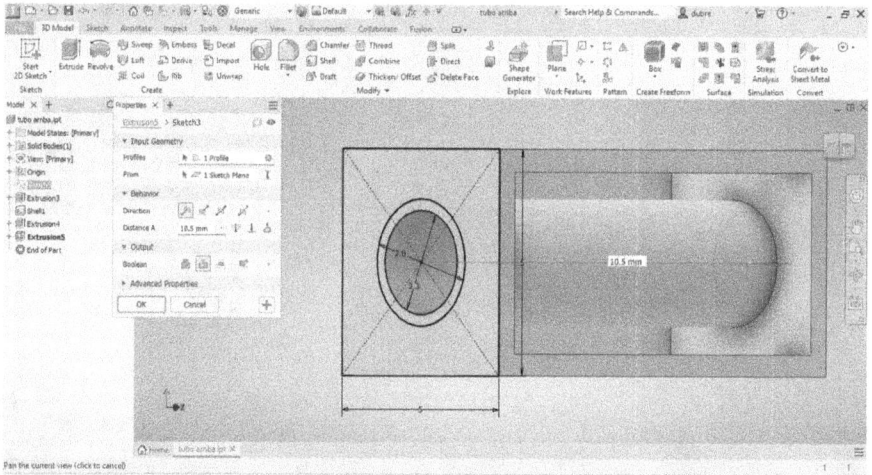

FIGURE 3.42
Extrusion of the internal face of the pipe.

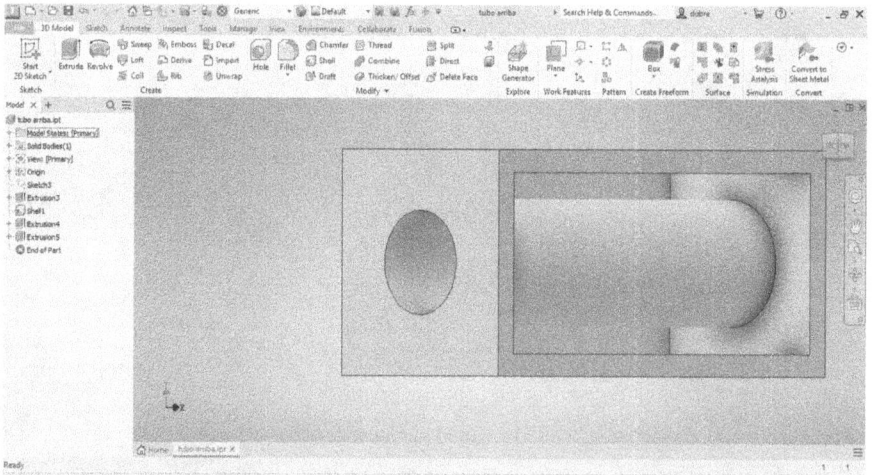

FIGURE 3.43
Finished structure.

3.3.3.4 Outlet Pipe

19. To start, repeat the first two steps of the previous circle monolith geometry. Sketch a rectangle starting from the origin of the plane by selecting the "Two point" option. It should measure 10.5 mm in length and 5 mm in height. Then trace a line to split it in half (lengthwise). From this guideline, draw two lines (10.5 mm in length) considering an angle of 45°. For more details, see Figure 3.44.

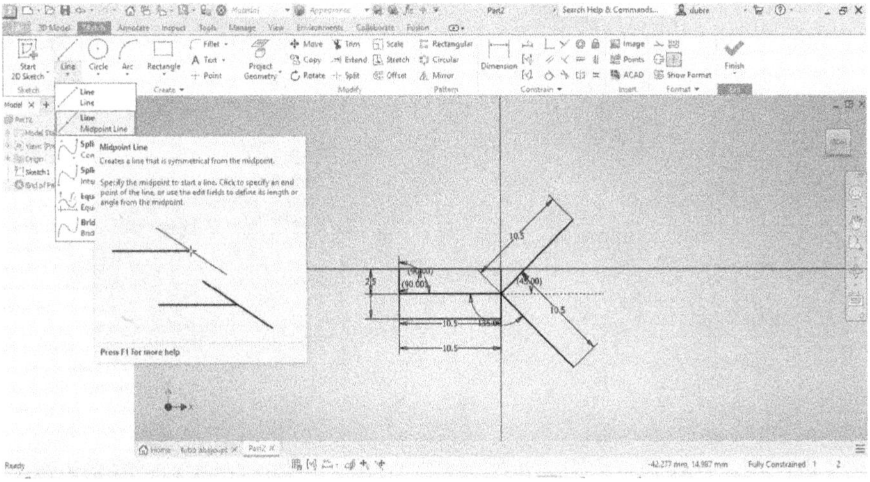

FIGURE 3.44
Initial sketch of the outlet pipe.

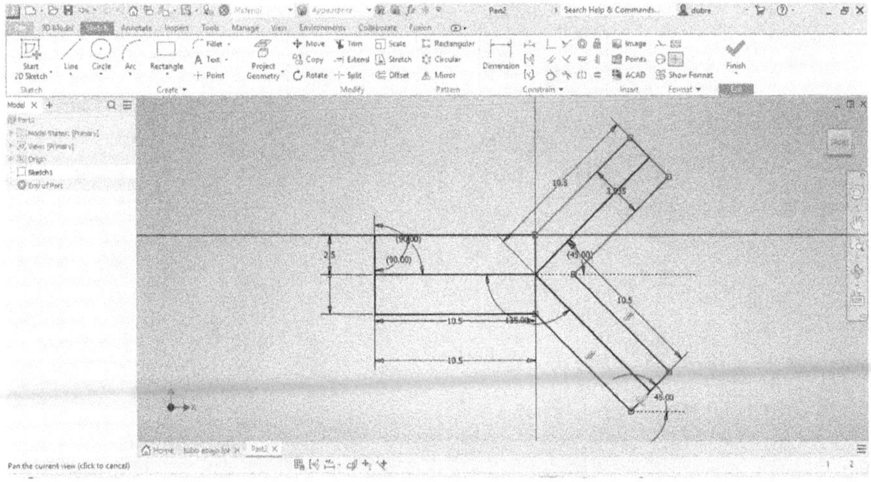

FIGURE 3.45
Complete sketch of the outlet pipe with the angles.

20. Continue drawing the walls of the Y-pipe. The final sketch should look like the one shown in Figure 3.45.

21. On "Sketch" tab, look for the "Fillet" the option and apply it to all the external edges generate on the Y-pipe. It has to measure 2 mm (Figure 3.46).

22. Click on the "3D Model" tab and go to the "Work Features" panel to create a new plane centering on the line that divides in half the

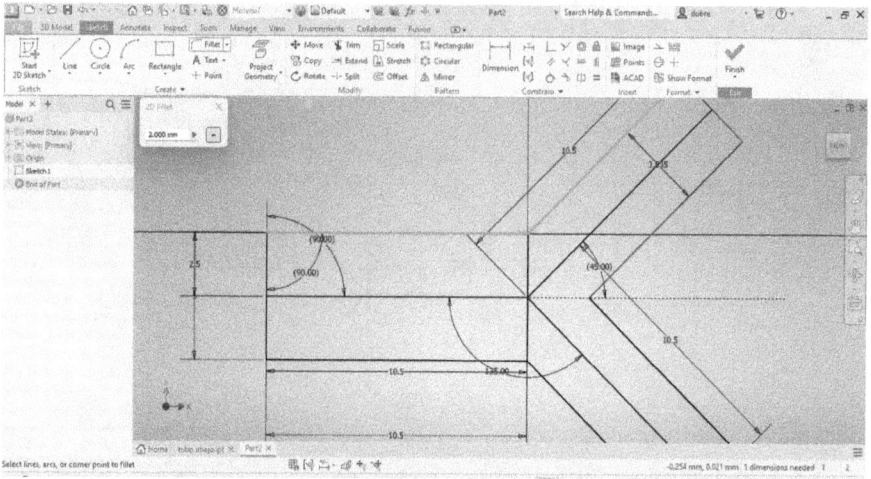

FIGURE 3.46
Joining the edges of the outlet pipeline.

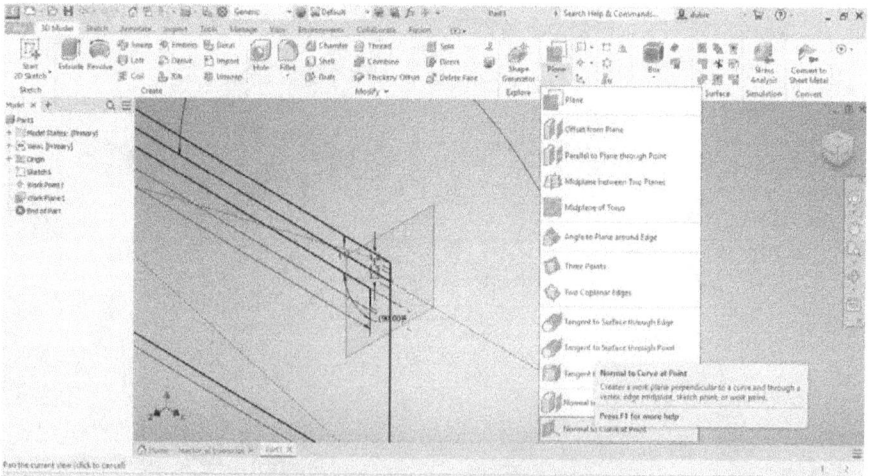

FIGURE 3.47
Creation of perpendicular planes to design the "Y" outlet of pipe.

initial rectangle and perpendicular to the XY plane, just as shown in Figure 3.47. Repeat it for each pipe terminal.

23. Draw two circles with a diameter of 3 mm and 2.8 diameter, respectively. Start from the center of the newly created planes. Again, repeat it for each pipe terminal. Then, proceed to extrude the external case surface. In this case, use the TWO DIRECTION, and it must measure 5 mm (Figure 3.48).

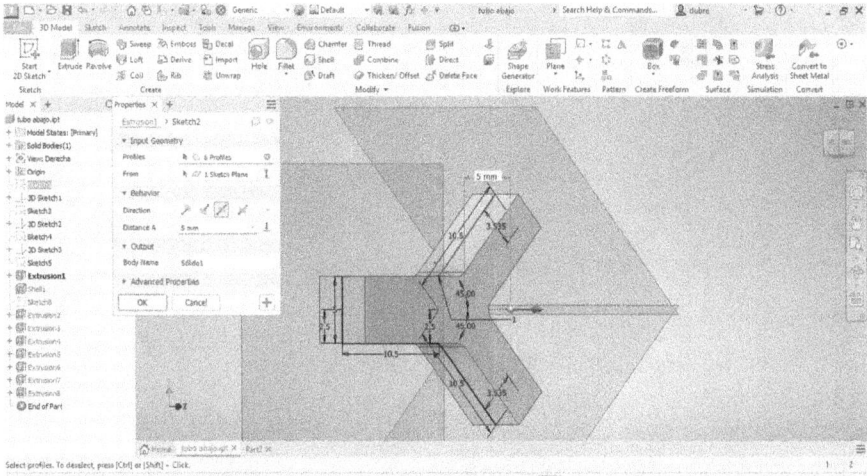

FIGURE 3.48
Extrusion of the external structure.

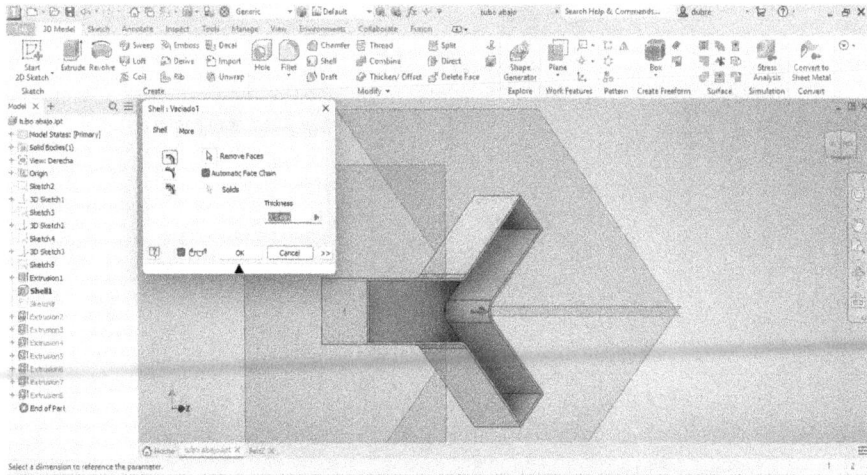

FIGURE 3.49
Shelling of the external structure.

24. In the same tab, select "Shell" (1 mm to the center) and remove the front face. The result should look like Figure 3.49.

25. Right click on "Sketch 1" showed on the right menu and choose "Share sketch," to continue editing the initial sketch. Now, extrude the 3-mm diameter circle, using the REVERSE direction; it must measure 10.5 mm (Figure 3.50).

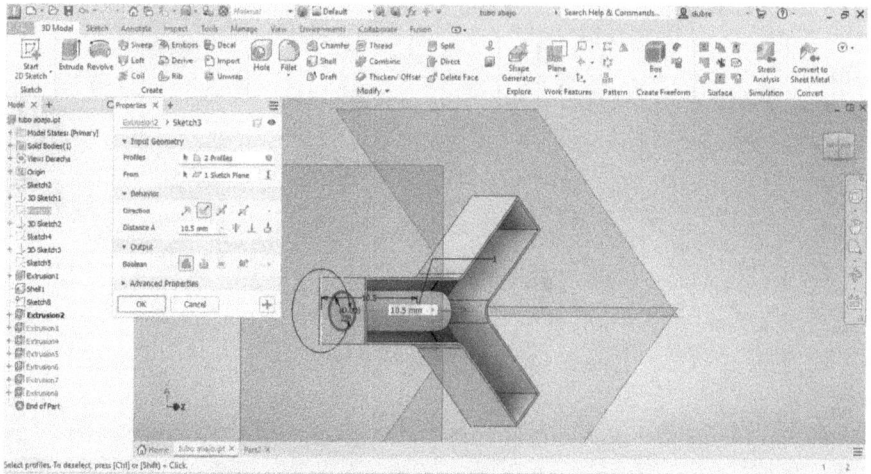

FIGURE 3.50
Extrusion of the external face of the pipe.

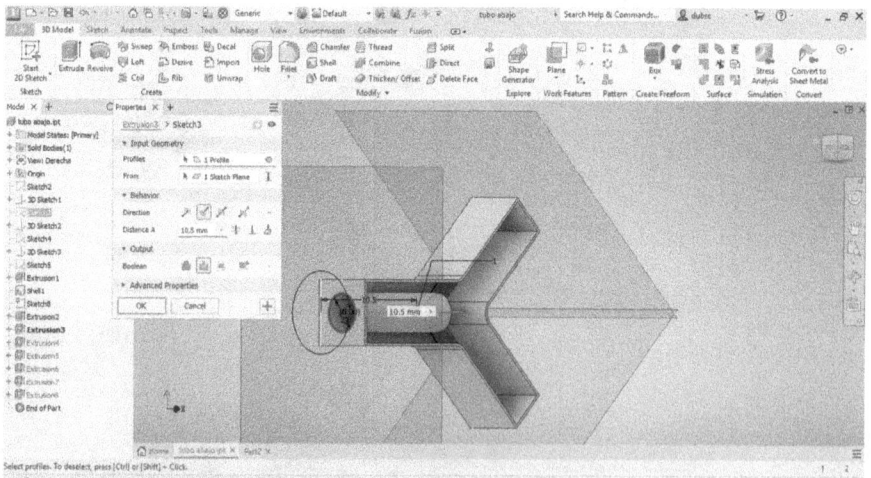

FIGURE 3.51
Extrusion of the internal face of the pipe.

26. Extrude the 2.8-mm diameter circle, using the DEFAULT direction. It must measure 10.5 mm and select CUT (Figure 3.51).
27. The resulting structure should look like Figure 3.52.

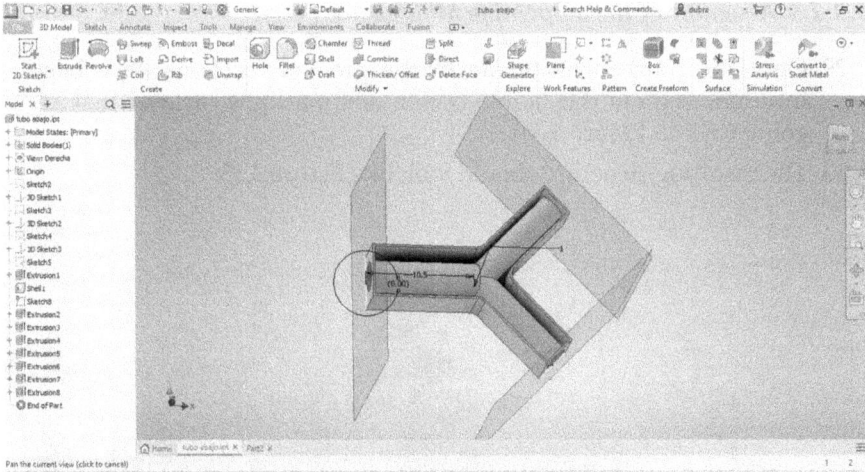

FIGURE 3.52
Finished structure.

28. Open a new project in Inventor® and select "Assemble (.iam)." Go to "Assemble" tab and click on "Place component" to open all the components of the serpentine reactor (Figure 3.53).

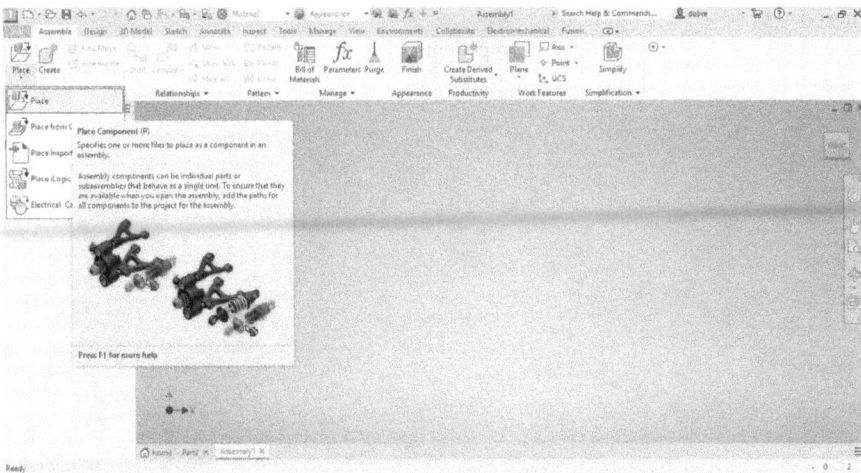

FIGURE 3.53
Insert serpentine millireactor components into the program environment.

3.3.3.5 Assembly of the Serpentine Millireactor Parts

29. Once the components made in the previous sections have been inserted, find out the "Joint" option and mark the surfaces that are going to link (Figure 3.54).

30. The resulting structure should look like Figure 3.55.

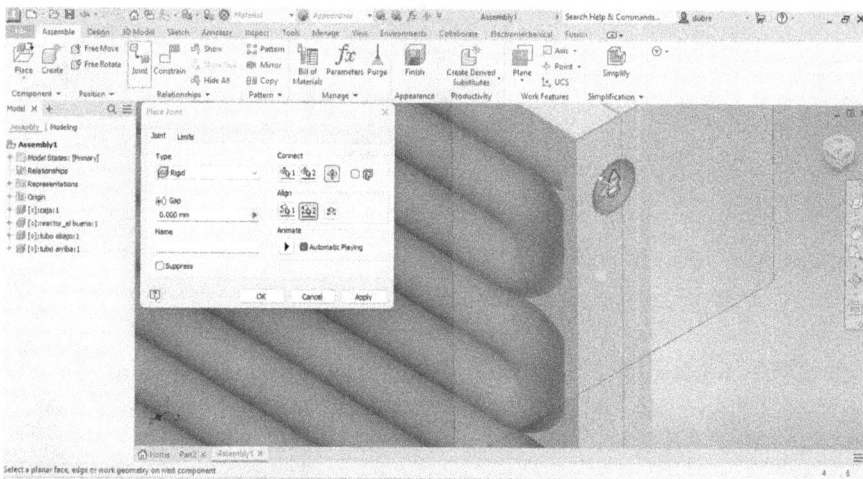

FIGURE 3.54
Assembly of serpentine millireactor components.

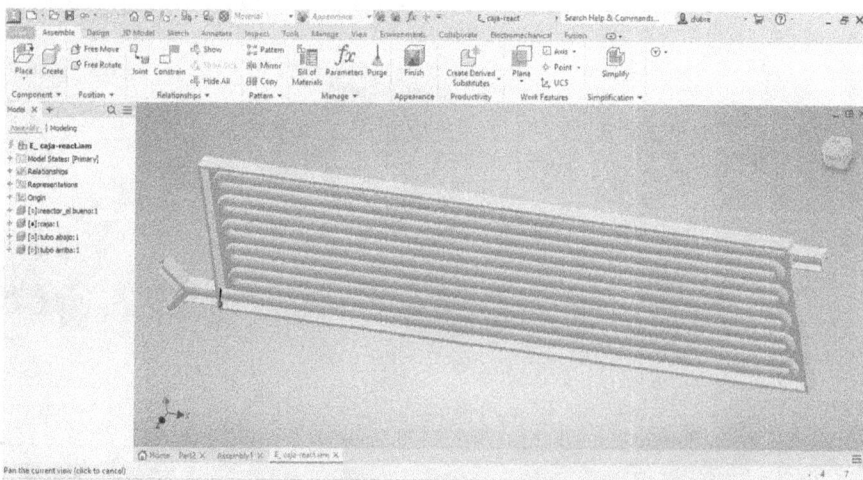

FIGURE 3.55
Serpentine reactor finished.

3.3.4 Student Challenge

To improve the 3D model skills, you can design a serpentine millireactor using heart shape. The constrictions and expansion in the heart shape will increase Reynolds number and improve the mixing of the biphasic solution of methanol and oil. You can find more details in the following datasheet: https://drive.google.com/file/d/1Wn2aOdDbcdFRxL_omPnznzHZACQaA7-v/view?usp=sharing

3.3.5 Remarks

Write down any comments you have regarding Practice 3:

_____

_____

_____

_____

_____

_____

4

Creating a Bowling Chatbot Using LangChain

Summary and Preliminary Questions

In this activity, you will collect and analyze data from a mini-bowling game to build a machine learning (ML) model. You will also explore how artificial intelligence (AI) tools, such as LangChain, can provide insights and recommendations based on gameplay data. The goal is to understand how structured data collection and analysis can improve decision-making and performance strategies.

Before starting, answer the following questions:

1. What is ML, and how can it be applied to sports analytics?

   _____

   _____

   _____

2. How does AI differ from ML?

   _____

   _____

   _____

3. What is LangChain, and how can it be used to create interactive applications?

   _____

   _____

   _____

   _____

   _____

   _____

DOI: 10.1201/9781003374503-4

4.1 Theoretical Framework

4.1.1 Understanding ML

ML is a subset of AI that allows computers to automatically learn patterns from data and improve over time without needing explicit programming. Rather than following a set of predefined rules, ML models use algorithms to find relationships between input data and outcomes. These patterns are used to make predictions or decisions. The power of ML lies in its ability to adapt and improve as it is exposed to more data, allowing for better accuracy in predictions over time. The three primary types of ML are **Supervised Learning**, **Unsupervised Learning**, and **Reinforcement Learning (RL)**, each serving different purposes and use cases.

- **Supervised Learning**: This is one of the most common types of ML, where the model is trained using labeled data. This means the data has both the input features (e.g., throw angle, speed) and the correct output or target value (e.g., number of pins knocked down). The model learns the relationship between the input and the output, so it can make accurate predictions on new, unseen data. Algorithms used in supervised learning include the following:
  - **Linear Regression**: Used for predicting continuous values, such as predicting the score in a bowling game based on certain variables like the angle of the throw.
  - **Decision Trees**: These are used for classification tasks, such as predicting whether a throw will result in a strike or not. They split the data into decision points based on the input features to reach a conclusion.
  - **Neural Networks**: These are complex algorithms that mimic the human brain's structure, enabling them to detect intricate patterns, even in large and complex datasets. They are particularly useful for tasks like image recognition, speech processing, and any application requiring deep pattern recognition.
- **Unsupervised Learning**: In contrast to supervised learning, unsupervised learning works with data that doesn't have labeled outcomes. The goal is to uncover hidden structures or patterns within the data without specific guidance on what to look for. Common algorithms include the following:
  - **Clustering**: This algorithm groups similar data points together. For example, clustering can be used to group similar bowling strategies based on performance, helping identify effective techniques or areas for improvement.

- **Principal Component Analysis (PCA)**: This is a dimensionality reduction technique, used to simplify a complex dataset while preserving the most important information. PCA can help reduce the number of features in a dataset, making it easier to analyze without losing significant detail.

- **Reinforcement Learning**: In RL, an agent learns by interacting with an environment through trial and error. The agent receives feedback in the form of rewards or penalties based on its actions, allowing it to learn over time which actions lead to better outcomes. This approach is commonly used in gaming, robotics, and optimization problems, where the system's goal is to maximize cumulative rewards through a series of decisions.

4.2 How to Pick the Best Model

When choosing the best ML model for a specific task, it's essential to consider several factors, including the dataset size, the type of prediction you're trying to make, and the computational resources available. Each type of model has its strengths and limitations, so selecting the right one requires balancing performance, complexity, and practicality.

1. **Dataset Size**: The size of your dataset plays a crucial role in deciding which model to use. If you have a small dataset, simpler models like **linear regression** or **decision trees** may work well since they don't require a large amount of data to train effectively. However, if you have a large dataset, more complex models like **neural networks** can take advantage of the vast amount of data to detect intricate patterns and produce better predictions.

2. **Type of Prediction**: The nature of the prediction you're making is another important factor. If you're predicting a **continuous outcome** (e.g., how many pins will fall in a bowling game), you'll typically use **regression models** like linear regression or more advanced models like **neural networks**. If you're classifying an event (e.g., determining whether a throw is "good" or "bad"), **classification models** such as decision trees or support vector machines (SVM) are more appropriate. The choice depends on whether you're predicting a numeric value or a category.

3. **Computational Power**: More advanced models, especially deep learning models like **neural networks**, can be computationally expensive and require powerful hardware (e.g., GPUs) for training. If you're working with limited computational resources, it's often better to start with simpler models that can run on standard

computers. Simpler models, like decision trees and linear regression, typically don't require as much power and can still deliver solid results, especially when your dataset is smaller or less complex.

In summary, the best model is not necessarily the most complex one, but the one that best fits your data and problem. Simpler models work well with smaller datasets and less computational power, while more complex models are suited for large, intricate datasets that require deep pattern recognition. The key is to understand your specific problem and constraints and select a model accordingly to achieve the best results efficiently.

4.3 What Is the Difference between AI and ML?

AI and ML are terms that are often used interchangeably, but they are not the same thing. While both deal with the development of intelligent systems that can perform tasks autonomously, they differ in scope, methodology, and application. Here's a breakdown of their key differences:

- **AI** is the broader concept. It refers to the creation of systems or machines that can mimic human intelligence to perform tasks such as reasoning, learning, problem-solving, language understanding, and even creativity. AI aims to simulate human cognitive abilities, making machines capable of performing complex tasks in ways that would typically require human intelligence. This includes everything from rule-based systems to more advanced, self-learning systems.

  Example: A chatbot that can understand user queries and respond appropriately is an example of AI. It doesn't necessarily need to learn from data or improve over time, but rather relies on programmed rules or logic to simulate intelligence.

- **ML** is a subset of AI. It focuses specifically on the idea that machines can learn from data, improve their performance over time, and make predictions or decisions based on patterns in the data, without being explicitly programmed for each specific task. ML uses algorithms and statistical models to analyze and draw insights from data. Rather than relying on human programmers to define every rule and action, ML allows systems to "learn" from historical data and improve their behavior automatically.

  Example: A recommendation system on a streaming platform (like Netflix) that learns from user behavior to suggest movies and TV shows is an example of ML. It doesn't operate based on hard-coded rules but rather improves its suggestions over time by analyzing user preferences and behaviors.

4.3.1 Key Differences

1. **Scope:**
   - AI is the broader field encompassing various methods to simulate intelligence in machines.
   - ML is a specific approach within AI that focuses on data-driven learning and prediction.

2. **Methodology:**
   - AI can include both **rule-based** systems, which follow predefined instructions, and learning-based systems (like ML), which adapt and improve from experience.
   - ML specifically involves learning from data through algorithms and statistical techniques.

3. **Goals:**
   - AI aims to create machines that can perform tasks that typically require human intelligence, such as speech recognition, decision-making, and problem-solving.
   - ML aims to enable machines to automatically improve their performance by learning patterns and insights from data.

4. **Applications:**
   - AI applications include robotics, natural language processing, expert systems, and general problem-solving systems.
   - ML applications are often more specific, such as predictive analytics, recommendation systems, image recognition, and speech recognition.

In summary, AI is the overarching field that encompasses a variety of techniques, including ML, to enable machines to simulate human-like intelligence. ML is a subset of AI focused on enabling systems to learn from data and make predictions or decisions without explicit programming. While AI is more about creating intelligent systems, ML is specifically about using data to improve these systems' capabilities over time.

4.4 Exploratory Data Analysis (EDA)

EDA is a crucial first step in any data science project, as it helps you understand the structure, patterns, and nuances of the dataset before applying any ML models. EDA involves a series of techniques aimed at summarizing the dataset's main characteristics, identifying potential issues, and gaining insights that can inform the choice of models and features for prediction.

The process involves several stages, including **data cleaning, data visualization, feature selection**, and **correlation analysis**.

- **Data Cleaning**: The first step in EDA is to ensure the dataset is clean and free from errors or inconsistencies. This involves:
  - **Checking for Missing Values**: Missing or incomplete data can negatively impact model performance. Techniques like imputation (filling in missing values with averages, medians, or predictions) or simply removing rows or columns with missing data are often applied.
  - **Identifying Duplicate Entries**: Duplicate records can distort analysis and model performance. Removing duplicates ensures that the dataset is accurate and doesn't overemphasize certain data points.
  - **Inconsistent Formatting**: Ensuring that all data is formatted consistently (e.g., units of measurement, date formats, or categorical labels) helps prevent errors during analysis. For example, throw speeds should be in the same units (miles per hour or kilometers per hour) across the dataset.
- **Data Visualization**: Visualizing the data is one of the most powerful ways to gain insights into its underlying patterns and relationships. Common visualization techniques include:
  - **Scatter Plots**: Scatter plots are often used to observe the relationship between two continuous variables. For instance, plotting the relationship between throw angle and the number of pins knocked down could help identify patterns, such as whether certain angles consistently lead to higher scores or strikes.
  - **Histograms**: Histograms are useful for understanding the distribution of a single variable, such as the distribution of bowling speeds. They allow you to see how the data is spread out and whether there are any skewed distributions or outliers.
  - **Box Plots**: These plots display the distribution of data through their quartiles and can help identify outliers or extreme values in a variable like bowling speed or pin count.
- **Feature Selection**: In any dataset, not all features are equally important for predicting the target variable. Feature selection helps identify the most relevant attributes (or features) that have the greatest impact on the outcome. For example, in a bowling dataset, features like **throw angle, speed**, and **spin** might have a significant effect on the number of pins knocked down, while other features (such as player height or weight) might have less relevance. By focusing on the most important features, you can reduce complexity, improve model performance, and make the model more interpretable.

- **Correlation Analysis**: Correlation analysis examines the relationships between variables to determine how strongly they are related. This helps identify which features are most important in predicting the target variable. For instance, if **throw speed** and **pin count** have a strong positive correlation, it might indicate that faster throws are more likely to knock down more pins. Similarly, if **throw angle** is strongly correlated with **strike rates**, adjusting throw angles could be a key factor in improving performance. Correlation matrices and heatmaps are often used to visualize the strength and direction (positive or negative) of these relationships. Identifying these relationships can guide feature engineering decisions and help refine the model.

In summary, EDA is a critical process that lays the foundation for building ML models. By cleaning the data, visualizing patterns, selecting important features, and analyzing correlations, you gain a deeper understanding of the dataset, which in turn leads to more informed decisions when building and training models. EDA helps to ensure that the data used is both relevant and high-quality, increasing the chances of building a successful predictive model.

4.5 Implementing LangChain for AI Suggestions

LangChain is a powerful framework designed to simplify the creation of AI applications that can interact with users in a conversational manner. It provides tools and abstractions to seamlessly combine various AI models, including large language models (LLMs), external data sources, APIs, and even custom logic to build sophisticated applications that can reason, fetch information, and adapt to dynamic user queries. It is particularly useful for developing conversational agents (chatbots) that can handle complex interactions by leveraging language models and chaining them together in sequences to perform advanced tasks. LangChain allows developers to create applications where AI doesn't just respond with pre-programmed answers but actively learns from interactions and offers personalized suggestions based on user input.

A key concept within LangChain is the idea of a **Chain**. A chain refers to a sequence of operations that are performed to handle a specific task. Each operation in the chain could involve processing input, interacting with an external tool or API, or leveraging an LLM to generate a response. Chains allow developers to control the flow of how the AI processes user input and produces output, making the system more dynamic and capable of handling a variety of tasks. For example, in the context of the bowling game, a chain might first parse the input data (pins knocked down, angle, speed) and then call an LLM to analyze patterns and generate suggestions.

An **Agent** in LangChain is a higher-level abstraction that combines multiple chains and can decide which chain to execute based on the context of the conversation. Agents are equipped with the ability to reason, make decisions, and choose appropriate actions based on the input they receive. They can perform tasks like querying external databases, managing conversational state, or combining multiple LLM outputs into a final response. In the case of the bowling chatbot, an agent would be responsible for deciding when to suggest changes in technique based on patterns it identifies in user inputs, ensuring the interaction feels fluid and intelligent.

In the context of AI models, a **Transformer** is an architecture used in deep learning models, such as GPT, that excels at processing sequences of data (like text) and understanding contextual relationships within them. Transformers leverage mechanisms like **attention** to focus on different parts of the input data when making predictions, allowing them to handle long-range dependencies and produce more coherent, contextually relevant outputs. This is crucial for tasks like language translation, summarization, and conversation, where understanding context is key to generating appropriate responses.

A **GPT** (Generative Pretrained Transformer) is a type of language model built on the Transformer architecture, and it is pre-trained on vast amounts of text data. It is capable of generating human-like text based on the prompts it receives. GPTs can perform a wide range of natural language processing tasks, such as answering questions, completing sentences, or suggesting ideas, making them ideal for conversational applications. With LangChain, GPTs can be integrated into chains and agents to provide context-aware, personalized responses that enhance the user experience. For example, GPT might analyze the input bowling data, identify patterns, and suggest improvements in the player's technique to optimize their performance.

In this activity, we will:

1. **Load the Dataset**: Convert bowling data into a structured format.

2. **Use a Pre-Trained LLM** to process and interpret inputs.

3. **Generate Strategy Suggestions**:
   - If a player frequently knocks down only a few pins, the chatbot may suggest adjusting throw speed.
   - If an angle consistently leads to strikes, the chatbot can highlight this pattern.

4. **Deploy the Chatbot**: Use **Streamlit, Gradio,** or the built-in functions in **Google Colab** to make the chatbot interactive.

Example User Interaction:

User: "I knocked down 5 pins with an angle of 30° and medium speed."
Chatbot: "Try increasing your angle to 35° for better accuracy."

4.6 Why Would You Use ML to Solve This Kind of Problem?

ML offers significant advantages when solving problems that involve large, complex datasets or tasks that are difficult to explicitly program with traditional rule-based methods. In the context of a problem like predicting or optimizing bowling performance, using ML can provide deeper insights, more accurate predictions, and personalized suggestions, all while adapting to changing patterns over time. Here are some key reasons why you would use ML to solve this kind of problem:

4.6.1 Ability to Identify Patterns in Complex Data

Bowling performance is influenced by multiple factors such as **throw angle**, **speed**, **spin**, and even external conditions like lane oil patterns or ball weight. The relationships between these variables and the outcome (e.g., number of pins knocked down) are complex and not always intuitive. Traditional programming approaches would require you to manually define all possible rules and outcomes, which would be time-consuming and prone to human error.

ML, on the other hand, can automatically detect hidden patterns in large datasets. By analyzing past throws and outcomes, an ML model can uncover subtle relationships—such as how slight changes in throw angle might improve accuracy—without needing explicit instructions. This allows the system to adapt and improve as it is exposed to more data, potentially discovering patterns that even expert bowlers might overlook.

4.6.2 Personalization and Adaptation

One of the most powerful aspects of ML is its ability to personalize recommendations based on individual performance. In the context of bowling, this means that the system could learn a player's unique strengths and weaknesses, tailoring suggestions specifically to them. For instance, if a player consistently knocks down only a few pins, the system might suggest specific adjustments (e.g., increasing speed or changing the angle), personalized to that player's unique style.

As the player continues to provide data—whether it's through more games or varying conditions—the model can adapt and refine its suggestions, ensuring that the advice stays relevant and effective over time.

4.6.3 Predictive Power

ML models are excellent at making predictions based on historical data. For example, once the model has learned from a series of past throws, it can predict the outcome of future ones under similar conditions. If a player inputs data like "I knocked down 7 pins with a throw angle of 20° and speed 15 mph," the system could predict whether this setup is likely to be successful and suggest adjustments.

Predictive capabilities also extend to things like scoring trends or performance over time. A player might get suggestions on whether they're improving or if they need to adjust their strategy based on past performance.

4.6.4 Handling Large, Multivariable Datasets

In real-world scenarios, bowling performance may depend on a wide range of factors. While humans can sometimes intuitively consider a few variables (e.g., speed and angle), ML algorithms are well-suited for handling large datasets with many variables and complex relationships. This means ML can simultaneously consider **throw speed, angle, spin**, and even subtle environmental factors (like lane conditions or ball type) to produce more accurate and holistic predictions. It can also weigh which variables are most important for making a successful throw, helping to refine performance optimization strategies.

4.6.5 Continuous Improvement

Unlike traditional approaches, where rules are fixed, ML models continue to improve over time as they are fed new data. As the system interacts with more data (from different players, different lanes, different conditions), it can evolve and refine its models to become more accurate. For example, if the system initially struggles with predicting performance on wet lanes but receives more data from such conditions, it can gradually improve its accuracy for those scenarios. This continuous learning allows ML-based systems to stay relevant and increasingly useful in dynamic environments.

4.6.6 Scalability

Once an ML model is trained and optimized, it can easily be scaled to handle larger datasets or applied to many different users. For example, the same bowling performance model could be used by many players, each with their own dataset, allowing the system to provide personalized advice to a wide audience. This scalability is harder to achieve with traditional rule-based systems, which would require manual updates and adjustments as the dataset grows or the scope of the problem changes.

4.6.7 Efficiency and Automation

ML automates the process of analyzing and making predictions based on large volumes of data, which can save significant time and effort compared to manual methods. For bowlers or coaches, this means they can receive automated insights and suggestions without having to perform all the calculations or analysis themselves. ML-based systems can instantly analyze past data, evaluate performance, and offer recommendations, making them highly efficient tools for improving performance in real-time or over the long term.

In short, **ML** is ideal for solving complex, data-driven problems like optimizing bowling performance because it can:

- Discover hidden patterns in data.
- Make accurate, personalized predictions and suggestions.
- Adapt and improve over time with more data.
- Handle large datasets with many variables.
- Continuously refine itself to offer better advice.
- Scale across multiple users, creating a more powerful system as more data becomes available.

Using ML in this context not only helps optimize a player's technique but also provides actionable insights that can lead to significant improvements, making it a valuable tool for anyone looking to enhance their bowling performance.

4.7 Multiple Choice Questions

1. What is the goal of ML?
   A. To write rules that a computer follows exactly
   B. To allow computers to learn patterns from data and make predictions
   C. To replace all human jobs with robots
   D. To store large amounts of data efficiently
2. What are the two main types of ML?
   A. Supervised and Unsupervised Learning
   B. Deep Learning and AI
   C. Classification and Regression
   D. Training and Testing
3. In supervised learning, what do we need in our dataset?
   A. Labels or correct answers for each example
   B. Only raw, unstructured data
   C. Data with no patterns
   D. A human to manually adjust every prediction

4. What is overfitting in ML?
   A. When a model is too simple and performs poorly
   B. When a model memorizes the training data but does not general-
      ize well
   C. When a model has too many correct predictions
   D. When a model is trained with too little data
5. What is a common way to measure the performance of a ML model?
   A. The number of features in the dataset
   B. The amount of time spent training
   C. The accuracy or error rate on test data
   D. The complexity of the code used

4.8 Short Answer Questions

1. What is the difference between supervised and unsupervised learning? Which should you use for this project?

   _____

   _____

   _____

   _____

   _____

2. Why is it important to split data into training and testing sets?

   _____

   _____

   _____

   _____

3. What are the advantages of using regression models for prediction?

   _____

   _____

   _____

   _____

4. How does LangChain enhance conversational AI?

_____

_____

_____

_____

5. What are some real-world applications of AI-based recommendations?

_____

_____

_____

_____

4.9 Materials

- Mini-bowling lane
- Game card for bowling
- Mobile phone or notebook for data recording
- Spreadsheet software (Excel or Google Sheets)
- Python (Jupyter Notebook or Google Colab)
- Libraries: Pandas, Matplotlib, Scikit-learn, LangChain
- OpenAI API Key **NOTE: NEVER SHARE OR PUBLISH YOUR API KEY**

4.10 Experimental Procedure

4.10.1 Data Collection

Set up the bowling pins and have each player go through the ten rounds of the game. During the bowling game, record the following variables for each throw:

- **Round Number:** The sequence of the throw in the game.
- **Number of Pins Knocked Down:** The outcome of each throw.

- **Throw Angle (If Measurable or Estimated)**: The angle at which the ball was rolled.
- **Bowling Speed**: Categorized as "Slow," "Medium," or "Fast."
- **Total Score**: The cumulative score at the end of the game.

4.10.1.1 Data Recording Format

Player	Round	Pins Knocked Down	Throw Angle	Speed	Total Score
John	1	8	30°	Medium	50
Sarah	1	7	25°	Slow	48

Try to play the game so that you can collect a wide variety of data points. Don't have each player bowl fast and with a small angle, for example. Try to collect a variety of angles and speeds so that the ML model will have a wider breath of data. It is ok if no clear trends emerge here, real data is messy!

4.10.2 Data Preprocessing and EDA

4.10.2.1 Data Cleaning

- Check for missing or inconsistent values. Standardize speed categories (convert "Med" to "Medium," etc.), and one-hot encode all categorical variables.
- Remove any unnecessary columns. For example, does the player's name impact the prediction? Does the round number? How about the angle? **(HINT: There isn't a right answer to this question, it is very dependent on your dataset)**

4.10.2.2 Exploratory Data Analysis

- Use Python's Pandas to load and clean the data.
- Generate scatter plots to observe the relationship between throw angle and pins knocked down.
- Create histograms for bowling speed distribution.
- Identify trends and patterns in scoring performance.
- Plot the data using a heat map and analyze the trends of the dataset. Perform PCA if necessary.

4.11 Short Answer Questions

1. Are any variables highly correlated with the number of pins knocked down? Which ones? (Note: it is ok if the answer is no. Real data doesn't always reveal a clean trend.)

_____

_____

_____

_____

4.11.1 ML Model Development

4.11.1.1 Steps to Build the Model

1. **Test Models**: Start with choosing at least five different ML models and test them. Choose the model that performs best for your dataset.
2. **Cross-Validated the Model**: Using the model that worked best, run a five-fold cross-validation to see how the R^2, RSME, and MAE change over different test-train splits.
3. **Evaluate Performance**: Plot the R^2, RSME, and MAE for each fold using a bar chart, scatter plot, or other graph.

4.12 Short Answer Questions

1. What model did you choose for your dataset? Why was this model chosen?

_____

_____

_____

_____

2. Look at the R^2, RSME, and MAE values for your cross-validation. Do you think your model is overfit, underfit, or neither? What can you do to prevent over and under fitting?

_____

_____

_____

_____

_____

_____

_____

_____

4.12.1 LangChain Chatbot Implementation

Using LangChain, develop a chatbot that suggests strategies based on collected data.

4.12.1.1 Steps to Implement

1. Load the dataset into a structured format.
2. Use a pre-trained **LLM** to interpret player inputs.
3. Generate recommendations based on previous successful throws.
4. Deploy using **Streamlit or Gradio** (if time permits).

4.12.1.2 Example Interaction

User: "I knocked down 5 pins with an angle of 30° and medium speed."
Chatbot: "Try increasing your angle to 35° for better accuracy."

4.12.1.3 Short Answer Questions

1. How does LangChain enhance conversational AI?

_____

_____

_____

_____

_____

_____

_____

_____

2. What are some real-world applications of AI-based recommendations?

_____

_____

_____

_____

_____

_____

_____

_____

_____

4.13 Results and Analysis

Use the space below to write an analysis of your ML model. Discuss the following points below:

1. Document findings from EDA and model evaluation.
2. Compare actual vs. predicted bowling outcomes.
3. Discuss insights gained from AI-generated recommendations.

_____

_____

_____

_____

_____

_____

_____

_____

_____

4.14 Conclusions

By structuring gameplay data and applying ML, we can identify key factors that influence bowling performance. Using AI-driven tools like LangChain, we can further enhance strategic decision-making through real-time recommendations.

4.15 Answer Key

Here is an example of code that completed the above task. Also, if you want to copy the **Google Colab code**, here is an open access link to an example. It is important to note that the code was written for a specific collected experimental dataset, and so the decisions made about what model was the best were specific to the dataset. There is more than one right answer to the assignment. What choices you make during EDA, which model is chosen, and how you choose to use LangChain are all dependent on your data and the results you want to see from the bot.

```
##Imports

# Import necessary libraries
!pip install shap -q
!pip install pickle-mixin -q
!pip install numpy -q
!pip install langchain -q
!pip install openai -q
!pip install langchain-community -q
!pip install langchain-openai -q

import pandas as pd
import numpy as np
from sklearn.model_selection import cross_val_score,
train_test_split

from sklearn.preprocessing import StandardScaler,
OneHotEncoder
from sklearn.compose import ColumnTransformer
from sklearn.pipeline import Pipeline

from sklearn.preprocessing import FunctionTransformer
from sklearn.preprocessing import PolynomialFeatures
import numpy as np
```

```
from sklearn.ensemble import StackingRegressor
from sklearn.linear_model import SGDRegressor
from sklearn.ensemble import RandomForestRegressor
from sklearn.ensemble import GradientBoostingRegressor
from sklearn.ensemble import AdaBoostRegressor
from sklearn.tree import DecisionTreeRegressor
from sklearn.neural_network import MLPRegressor
from sklearn.svm import SVR
from xgboost import XGBRegressor
from sklearn.linear_model import LinearRegression
from sklearn.ensemble import ExtraTreesRegressor
import joblib
import seaborn as sns
import matplotlib
import matplotlib.pyplot as plt
import matplotlib as mpl
import scipy.stats as stats
import plotly.graph_objects as go
import plotly.express as px
from scipy.stats import spearmanr
from scipy.cluster import hierarchy
from scipy.spatial.distance import squareform
import pickle

from langchain.prompts import PromptTemplate
from langchain.agents import initialize_agent, Tool, AgentType
from langchain.agents import AgentExecutor
from langchain.memory import ConversationBufferMemory

# Import filters to remove unnecessary warnings
from warnings import simplefilter
import warnings
warnings.filterwarnings("ignore")
from sklearn.exceptions import ConvergenceWarning

from scipy.cluster import hierarchy
from scipy.spatial.distance import squareform

# Import filters to remove unnecessary warnings
from warnings import simplefilter
import warnings
warnings.filterwarnings("ignore")

from sklearn.metrics import mean_absolute_percentage_error,
mean_squared_error, r2_score, mean_absolute_error
import shap

import pickle
import openai
from langchain.prompts import PromptTemplate
```

```
from langchain.chat_models import ChatOpenAI
from langchain.chains import LLMChain
import re
from IPython.display import display, Markdown
```

Step 1: Preprocessing

To make an ML model, you need a dataset in order to train the model. For this project, we collected a dataset at Hooplas of mini bowling.

```
import pandas as pd

# Replace '/path/to/your/file.xlsx' with the actual file path
file_path = '/content/hooplas_dataset.xlsx'

# Read the Excel file into a Pandas DataFrame
df = pd.read_excel(file_path)

# Print the first few rows of the DataFrame
print("The first 5 rows of the dataset:")
print(df.head())

# Print the description of the DataFrame
print("The descriptive statistics of the dataset")
print(df.describe())

# Get the data types of all columns
data_types = df.dtypes
# Print the datatypes of each column
print("The data types of each column:")
print(data_types)

# Get the numeric columns
numeric_cols = df.select_dtypes(include=['int64', 'float64']).
columns
# Get the categorical columns
categorical_cols = df.select_dtypes(include=['object']).
columns
# Print the results
print("Numeric columns:", numeric_cols)
print("Categorical columns:", categorical_cols)
```

We can drop irrelevant information from the dataset, which includes the round number and the player name. This will also allow the final dataset to not have any personal information about the people in the dataset. We can also remove the total score of each of the players because we are just focusing on the per round scores for the target of the ML model. This will leave only the features (speed and angle) and target (pins knocked).

```
df = df.drop(['Round','Player_Name','Total_Score'],axis=1)
```

Next, we can check for missing values in the dataset. When creating a ML model, it is very important that your columns don't contain missing data because that can skew the model and make it less accurate.

```
import pandas as pd
import numpy as np

missing_data = df.isnull().sum()
print("Total missing values:")
print(missing_data)
```

This dataset has no missing data, so we don't need to worry about any missing data handling. This makes sense because we collected all of the datapoints individually and made sure not to miss any data points when collecting the data as to ensure the highest accuracy of the set.

Next, we can check for outliers. It is important to handle outliers because they can also skew any ML models that are derived from the dataset. So, we first need to locate the outliers. If there are outliers, we can decide what to do with them based on the quantity of outliers in the dataset.

```
import matplotlib.pyplot as plt
import numpy as np

# Calculate outliers for each column
outliers_dict = {}

# Only check the columns that are numerical and would contain
outliers
columns_to_check = df.select_dtypes(include=['int64',
'float64']).columns

# Create boxplots for specified columns
df[columns_to_check].boxplot()
plt.title("Boxplots for Numerical Columns")
plt.ylabel("Values")
plt.xticks(rotation=45)
plt.show()

# Check for outliers and plot the boxplot
for col in columns_to_check:
    q1 = np.quantile(df[col], 0.25)
    q3 = np.quantile(df[col], 0.75)
    iqr = q3 - q1
    lower_bound = q1 - 1.5 * iqr
    upper_bound = q3 + 1.5 * iqr
```

```
  outliers = df[col][(df[col] < lower_bound) | (df[col] >
upper_bound)]
  outliers_dict[col] = outliers.tolist()

# Print outliers for each column
for col, outliers in outliers_dict.items():
  if outliers:
      print(f"Outliers in column '{col}': {outliers}")
  else:
      print(f"No outliers found in column '{col}'")
```

There are outliers in this dataset, but there is also not a lot of data points. Removing the outliers could remove too large of a portion of the dataset and make the ML model that we are writing inaccurate, so for now the outliers can stay in the set.

Next, we need to one-hot encode the categorical variables in the dataframe. This will allow a ML model to handle the categorical variables by transforming them into binary vectors where if the category is true, the value is 1 and if it is false the value is 0.

```
# Perform one-hot encoding
df = pd.get_dummies(df, columns=['Speed'], prefix='',
prefix_sep='')

# Ensure all NaN values are replaced with 0
df = df.fillna(0).astype(int)
print(df)
```

Finally, we can normalize the data frame. There are a lot of different numerical categories in this dataframe, and making sure that they are normalized will improve the accuracy of the model and prevent it from being skewed by numbers that are very different in magnitude.

```
from sklearn.preprocessing import MinMaxScaler
import pandas as pd

# Assuming 'df' is your DataFrame and you want to normalize
all columns
scaler = MinMaxScaler()

# Normalize the DataFrame
df = pd.DataFrame(scaler.fit_transform(df), columns=df.
columns)

# Display the normalized DataFrame
print(df)
```

Finally, we can export the normalized, encoded, and cleaned dataset to a csv file. This way if someone is trying to perform ML on the model they don't need to redo all of the steps above and can start with a dataset that is ready for modeling.

```
# Export DataFrame to CSV
csv_filename = 'hooplas_cleaned.csv' # Specify your file name
df.to_csv(csv_filename, index=False) # Use index=False to
exclude row indices from the CSV

print(f"DataFrame exported successfully to {csv_filename}")

## Step 2: Exploratory Data Analysis
---
```

To understand how to train the ML model, we need to look at the different trends that are present in the dataset. This will allow for further understanding of the dataset and so that we can create a better ML model.

To better understand the trends of the data, we want to then look at the pairplots to better understand if there are variables that relate strongly with one another. This will be helpful when we need to go through and reduce the dataset to make it more fit for ML analysis. Further, this can help us isolate which variables to target as the strongest correlations in the dataset.

```
import seaborn as sns
sns.pairplot(df)

# Set the style for seaborn plots
sns.set(style="whitegrid")

# Create a figure with multiple subplots
fig, axes = plt.subplots(1, 2, figsize=(18, 6))

# Histogram for normalized Throw_Angle (Bins range from 0 to 1)
axes[0].hist(df['Throw_Angle '], bins=10, edgecolor='black',
color='lightgreen')
axes[0].set_title('Histogram of Normalized Angle')
axes[0].set_xlabel('Normalized Angle')
axes[0].set_ylabel('Frequency')

# Histogram for normalized Pins_Knocked (Bins range from 0 to 1)
axes[1].hist(df['Pins_Knocked '], bins=10, edgecolor='black',
color='salmon')
axes[1].set_title('Histogram of Normalized Pins Knocked')
axes[1].set_xlabel('Normalized Pins Knocked')
axes[1].set_ylabel('Frequency')

# Adjust layout
plt.tight_layout()
# Show the histograms
plt.show()
```

Looking at the distributions of normalized angles and number of pins knocked, we can also see that the players preferred lower angles, yet the distribution for the number of pins knocked was relatively normal in shape.

```
import pandas as pd
import seaborn as sns
import matplotlib.pyplot as plt

# Calculate the correlation matrix
correlation_matrix = df.corr()

# Print the correlation matrix
print(correlation_matrix)

# Create a heatmap to visualize correlations
plt.figure(figsize=(12, 8))
sns.heatmap(correlation_matrix, annot=True, cmap='coolwarm')
plt.title('Correlation Matrix')
plt.show()

# Analyze significant correlations
significant_correlations = correlation_matrix[(correlation_
matrix > 0.7) | (correlation_matrix < -0.7)]
significant_correlations = significant_correlations.stack().
reset_index()
significant_correlations.columns = ['feature1', 'feature2',
'correlation']
print(significant_correlations)
```

Based on this, there are no strong trends between the number of pins knocked down and the throw speed when the speed is fast or medium this dataset. However, there is a significant trend between the throw angle and the number of pins knocked down in the dataset. Higher angles are correlated with lower scores.

```
## Step 3: Machine Learning Model
---
```

To determine the best ML model for predicting the number of pins knocked down in bowling based on throw speed and angle, we tested several different models. We used a variety of algorithms, each with unique characteristics, to evaluate their performance on our dataset. The models we tested included **Random Forest**, **Gradient Boosting**, **MLP (Multi-layer Perceptron)**, **SVR (Support Vector Regressor)**, and **Linear Regression**. By training and testing each model on the same dataset, we were able to compare their performance in terms of prediction accuracy. This process helped us identify which model achieved the best results and was most suited for our specific task, ensuring that we chose the most accurate and reliable model for bowling prediction.

```
from sklearn.model_selection import KFold
from sklearn.metrics import mean_squared_error, mean_absolute_
error, r2_score

# Split the dataset into features and target
X = df.drop('Pins_Knocked ', axis=1)
Y = df['Pins_Knocked ']

# Define the number of folds for K-Fold cross-validation
n_folds = 6

# Initialize empty lists to store evaluation metrics
rmse_scores = []
mae_scores = []
r2_test_scores = []
r2_train_scores = []

# Define the models dictionary
models = {
    'Random Forest': RandomForestRegressor(),
    'Gradient Boosting': GradientBoostingRegressor(),
    'MLP': MLPRegressor(),
    'SVR': SVR(),
    'Linear Regression': LinearRegression(),
}

# K-Fold cross-validation loop
kf = KFold(n_splits=n_folds, shuffle=True, random_state=42)
for name, model in models.items():

    # Loop through each fold
    for train_index, test_index in kf.split(X):
        X_train, X_test = X.iloc[train_index], X.iloc[test_index]
        y_train, y_test = Y.iloc[train_index], Y.iloc[test_index]

        # Train the model on the training data for this fold
        model.fit(X_train, y_train)

        # Predict on the testing data for this fold
        y_pred = model.predict(X_test)

        # Calculate evaluation metrics
        rmse = mean_squared_error(y_test, y_pred, squared=False)
# Calculate RMSE directly
        mae = mean_absolute_error(y_test, y_pred)
        r2_test = r2_score(y_test, y_pred)

        # Additionally, calculate R-squared on the training
data for each fold (optional)
        y_train_pred = model.predict(X_train)
        r2_train = r2_score(y_train, y_train_pred)
```

```python
        # Calculate R^2 difference
        r2_diff = r2_train - r2_test

        # Append the scores to the lists
        rmse_scores.append(rmse)
        mae_scores.append(mae)
        r2_test_scores.append(r2_test)
        r2_train_scores.append(r2_train)

    # Print average scores after all folds for each model
    print(f"{name}:")
    print(f"  Average RMSE: {np.mean(rmse_scores):.3f}")
    print(f"  Average MAE: {np.mean(mae_scores):.3f}")
    print(f"  Average R² Test Score:
{np.mean(r2_test_scores):.3f}")
    print(f"  Average R² Train Score:
{np.mean(r2_train_scores):.3f}")
    print(' ')

# Create a DataFrame to store the results
results_df = pd.DataFrame({
    'Model': ['Random Forest', 'Random Forest', 'Random
Forest', 'Random Forest', 'Random Forest', 'Random Forest',
             'Gradient Boosting', 'Gradient Boosting',
'Gradient Boosting', 'Gradient Boosting', 'Gradient Boosting',
'Gradient Boosting',

             'MLP', 'MLP', 'MLP', 'MLP', 'MLP', 'MLP',
             'SVR', 'SVR', 'SVR', 'SVR', 'SVR', 'SVR',

             'Linear Regression', 'Linear Regression', 'Linear
Regression', 'Linear Regression', 'Linear Regression', 'Linear
Regression'],
    'R^2 Test': r2_test_scores,
    'R^2 Train': r2_train_scores,
})

# Select the best model based on Test R², or lowest RMSE if R²
is identical
best_model_row = results_df.loc[results_df['R^2 Test'].
idxmax()]

print("\nBest model based on Test R² performance:")
print(best_model_row)

# Assuming 'results_df' is the DataFrame containing the model
performance metrics
models = results_df['Model']
train_r2 = results_df['R^2 Train']
test_r2 = results_df['R^2 Test']
```

```
# Melt the DataFrame for easier plotting
melted_df = results_df.melt(id_vars='Model', var_
name='Metric', value_name='R^2')

# Create a boxplot
plt.figure(figsize=(10, 6))
sns.boxplot(x='Model', y='R^2', hue='Metric', data=melted_df)
plt.title('R^2 Scores for Different Models')
plt.xlabel('Model')
plt.ylabel('R^2')
plt.legend(title='Metric')
plt.show()
```

According to the above analysis, the best model (the model with the best R^2 value, RSME value, and MAE value) will be linear regression. So, going forward, because we want the most accurate ML model possible, we will be using a linear regression model.

None of the models show a very good R^2, RSME, or MAE. This is because, after looking at the EDA, we can see that there isn't a very strong correlation between the score achieved and the speed or throw angle. This means that there are probably other variables influencing the score that we didn't measure. For now, we cannot collect more data and so we should move on with the best model.

Running this multiple times, different models had different R^2, RSME, MAE based on the test-train split. They are all not great because the data doesn't have strong correlations. However, linear regression seems to have a consistently ok R^2 value and so we will move forward with that.

```
# Initialize the model
model = LinearRegression()

# Initialize lists to store metrics
rmse_scores = []
mae_scores = []
r2_test_scores = []
r2_train_scores = []

# Define K-Fold parameters
n_folds = 5 # Adjust this to the number of folds you want
kf = KFold(n_splits=n_folds, shuffle=True, random_state=42)

# Loop through each fold
for fold, (train_index, test_index) in enumerate(kf.split(X), 1):

    X_train, X_test = X.iloc[train_index], X.iloc[test_index]
    y_train, y_test = Y.iloc[train_index], Y.iloc[test_index]

    # Train the model on the training data for this fold
    model.fit(X_train, y_train)
```

```
    # Predict on the testing data for this fold
    y_pred = model.predict(X_test)

    # Calculate evaluation metrics
    rmse = mean_squared_error(y_test, y_pred, squared=False)   #
Calculate RMSE directly
    mae = mean_absolute_error(y_test, y_pred)
    r2_test = r2_score(y_test, y_pred)

    # Additionally, calculate R-squared on the training data
for each fold
    y_train_pred = model.predict(X_train)
    r2_train = r2_score(y_train, y_train_pred)

    # Append the scores to the lists
    rmse_scores.append(rmse)
    mae_scores.append(mae)
    r2_test_scores.append(r2_test)
    r2_train_scores.append(r2_train)

    # Print the scores for each fold
    print(f"Fold {fold}:")
    print(f"  R² (train): {r2_train:.4f}")
    print(f"  R² (test): {r2_test:.4f}")
    print(f"  RMSE: {rmse:.4f}")
    print(f"  MAE: {mae:.4f}")
    print("-" * 30)

# Now calculate and print the average scores across all folds
avg_rmse = sum(rmse_scores) / n_folds
avg_mae = sum(mae_scores) / n_folds
avg_r2_test = sum(r2_test_scores) / n_folds
avg_r2_train = sum(r2_train_scores) / n_folds

print("\nAverage Scores Across All Folds:")
print(f" Average R² (train): {avg_r2_train:.4f}")
print(f" Average R² (test): {avg_r2_test:.4f}")
print(f" Average RMSE: {avg_rmse:.4f}")
print(f" Average MAE: {avg_mae:.4f}")

# Check if the number of scores matches the number of folds
assert len(rmse_scores) == n_folds, f"Expected {n_folds} RMSE
values, but got {len(rmse_scores)}"
assert len(mae_scores) == n_folds, f"Expected {n_folds} MAE
values, but got {len(mae_scores)}"
assert len(r2_test_scores) == n_folds, f"Expected {n_folds} R2
test values, but got {len(r2_test_scores)}"
assert len(r2_train_scores) == n_folds, f"Expected {n_folds}
R2 train values, but got {len(r2_train_scores)}"
# Now plot the metrics across the different folds
```

```
# Set up the figure size
plt.figure(figsize=(15, 5))

# Plot RMSE across the folds
plt.subplot(1, 3, 1)
plt.plot(range(1, n_folds + 1), rmse_scores, marker='o',
color='blue', label='RMSE')
plt.title('RMSE Across Folds')
plt.xlabel('Fold')
plt.ylabel('RMSE')
plt.xticks(range(1, n_folds + 1))
plt.grid(True)

# Plot MAE across the folds
plt.subplot(1, 3, 2)
plt.plot(range(1, n_folds + 1), mae_scores, marker='o',
color='red', label='MAE')
plt.title('MAE Across Folds')
plt.xlabel('Fold')
plt.ylabel('MAE')
plt.xticks(range(1, n_folds + 1))
plt.grid(True)

# Plot R-squared (test) across the folds
plt.subplot(1, 3, 3)
plt.plot(range(1, n_folds + 1), r2_test_scores, marker='o',
color='green', label='R2 Test')
plt.plot(range(1, n_folds + 1), r2_train_scores, marker='o',
color='orange', label='R2 Train')
plt.title('R-squared Across Folds')
plt.xlabel('Fold')
plt.ylabel('R-squared')
plt.xticks(range(1, n_folds + 1))
plt.grid(True)
plt.legend()

# Display the plots
plt.tight_layout()
plt.show()

# Specify the file name for the pickle file
pickle_filename = 'best_model.pkl'

# Open the file in write-binary mode and save the model
with open(pickle_filename, 'wb') as file:
pickle.dump(model, file)

print(f"Model has been saved to {pickle_filename}")
```

Step 4: Export to LangChain

In this step, the ML model's outputs and processes are integrated into LangChain, a framework designed to manage and orchestrate language model applications. LangChain facilitates seamless interaction between the model and external tools, databases, or APIs by providing a structured environment for handling language model-driven tasks. Exporting to LangChain enables the model to be used within complex workflows, such as chaining together multiple models, interacting with data sources, or managing conversational contexts. This step ensures that the ML model can be deployed in a flexible, scalable, and production-ready system, allowing for improved task automation, user interaction, and decision-making capabilities.

```
import pickle
import openai
from langchain.prompts import PromptTemplate
from langchain.chat_models import ChatOpenAI
from langchain.chains import LLMChain
import re
from IPython.display import display, Markdown

# Set your OpenAI API key
# Add your personal API key here. The API key used during
testing has been removed to
# ensure that private information of the owner is not leaked.
openai.api_key = 'ADD KEY HERE'

# Load your pre-trained model (Linear Regression model)
with open('best_model.pkl', 'rb') as f:
    best_model = pickle.load(f)

# Using the normalization parameters used during training
mean_angle = 15
std_angle = 5
mean_pins = 5
std_pins = 2

# Initialize OpenAI Chat-based model using LangChain
chat_model = ChatOpenAI(model="gpt-4", api_key=openai.api_key)

# Create a prompt template for your chatbot
template = """
```

You are a helpful assistant who provides strategies based on user inputs about a game where the user knocks down pins.
    The user provides:

- Speed ("fast" or "medium")
- Angle (between 0° and 30°)
- Number of pins knocked (between 0 and 10)

Based on the data, you should generate a suggestion for improving the user's performance if the actual score is lower than the predicted score. Your suggestions should be based on adjusting the speed or angle.

```
User Input: "{user_input}"

Predicted Pins: {predicted_pins}
Actual Pins: {actual_pins}

Generate a suggestion to improve the score.
"""

# Define the prompt template
prompt = PromptTemplate(input_variables=["user_input",
"predicted_pins", "actual_pins"], template=template)

# Set up the chain to use the model with the prompt
llm_chain = LLMChain(llm=chat_model, prompt=prompt)

# Function to extract information from natural language input
def extract_game_data(user_input):
    # Use regex to extract speed, angle, and pins knocked from
the input text
    speed_match = re.search(r"(fast|medium)", user_input,
re.IGNORECASE)
    angle_match = re.search(r"(\d{1,2})°", user_input)
    pins_match = re.search(r"(\d) pins", user_input)

    # If all matches are found, return the data
    if speed_match and angle_match and pins_match:
        speed = speed_match.group(1).lower()
        angle = int(angle_match.group(1))
        pins = int(pins_match.group(1))
        return speed, angle, pins
    else:
        raise ValueError("Could not extract valid data from the
input.")

# Function to denormalize the predicted pins
def denormalize_pins(predicted_pins):
    # Denormalize using the appropriate method (e.g., z-score
normalization)
    return predicted_pins * std_pins + mean_pins

# Define a function to get suggestions based on user input
def get_suggestion(user_input):
    try:
        speed, angle, pins = extract_game_data(user_input)
    except ValueError as e:
        return str(e)
```

```
    # Convert speed to one-hot encoded vector (e.g., [1, 0] for
'fast', [0, 1] for 'medium')
    speed_vector = [1, 0] if speed == 'fast' else [0, 1]

    # Prepare the data for the model (inputs: speed, angle)
    features = speed_vector + [angle]  # Only speed and angle
as features

    # Predict the number of pins knocked using the model
    try:
        predicted_pins_normalized = best_model.
predict([features])[0]  # Get the normalized predicted number
of pins
        predicted_pins = denormalize_pins(predicted_pins_
normalized)  # Denormalize the prediction
    except Exception as e:

        return f"Error in prediction: {e}"

    # Generate a suggestion using OpenAI if the score is lower
than predicted
    suggestion = None
    if pins < predicted_pins:
        response = llm_chain.run(user_input=user_input,
predicted_pins=predicted_pins, actual_pins=pins)
        suggestion = response

    # Format the result for Markdown output
    formatted_response = f"""
### Your Input:
**Speed**: {speed}
**Angle**: {angle}°
**Pins Knocked**: {pins}

### Predicted Pins (denormalized): {predicted_pins:.2f}

### Suggested Improvement:
{suggestion if suggestion else "Your actual score matches
or exceeds the predicted score."}

    ---
    Try experimenting with this suggestion and see how your
score improves!
    """

    # Display the formatted response using Markdown
    display(Markdown(formatted_response))

# Interactive loop to ask for game input
while True:
```

```
    user_input = input("Enter your game details (e.g., 'I
knocked down 5 pins with an angle of 30° and speed fast'): ")

    # Break out of the loop if the user types 'exit'
    if user_input.lower() == 'exit':
        print("Exiting chatbot.")
        break

    # Get a suggestion from the model
    print(get_suggestion(user_input))
    print("\nType 'exit' to end the conversation.\n")
```

FINAL CONCLUSIONS

The analysis of this data shows that in trying to improve your per round pins knocked, you should focus on lowering your angle to be as close to zero as possible. Lower release angles were correlated with higher number of pins knocked per round.

Further, this project shows that a ML model can be created that predicts the number of pins knocked based on the throw angle and the speed of the throw. A gradient boosting regression model was chosen because it has the best combined R^2, RSME, and MAE values. Although it is important to note that there wasn't a strong correlation in the dataset between the variables, and so the R^2 of the model is not particularly high across the test-train split.

The model is not perfectly accurate and doesn't have a high test-train R^2 split. This is because we have a really small dataset and the dataset wasn't collected with a very accurate sampling method (we just guessed the angle and speed). In the future, more data being collected can help to improve the model.

Finally, LangChain can be used in order to create a chatbot that can use this ML model to answer user questions about their bowling techniques. LLMs can be used to process natural language while the model can be used to make predictions.

5

Renewable Hydrogen Production

Summary

This chapter describes hydrogen production through renewable means. The topics discussed range from hydrolysis, electrolysis using solar energy, and dark fermentation. For hydrolysis, waste aluminum from soda, which could be recycled, is used. Electrolysis features a rig built to split water using proton exchange membrane (PEM) electrolyzer and then combine the gases using PEM fuel cells to produce energy that could power an electrical motor. This chapter contains student's tasks and safety guidelines to enable the students carry out the experiments on hydrogen production. However, fuel cells or microbial fuel cells are not discussed in this book. The hydrogen production by dark fermentation using wastewater sludge, juice as a carbon source, and an inoculum from sewage sludge is presented in this chapter. A step-by-step experimental guideline to help students or teachers replicate the experiments at home or in the school's laboratory is presented. The annex of this chapter contains a short video demonstration of hydrolysis, electrolysis, and fermentation process.

5.1 General Objectives of the Experiments

At the end of this chapter, the students are expected to learn the following:

1. Understand the theoretical framework of renewable hydrogen production.
2. Carry out experiments to produce hydrogen via hydrolysis, electrolysis and fermentation.
3. Identify different types of electrolyzers and their characteristics.
4. Analyze the experiments and determine the efficiency of the processes (Figure 5.1).

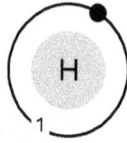

FIGURE 5.1
Representative of a hydrogen atom.

5.2 Hydrogen Production from Aluminum Hydrolysis

5.2.1 Summary and Preliminary Questionnaire

Hydrolysis is a two-syllable word hydro (water) and lysis (split) which refers to reactions that use water molecule to breakdown chemical bonds. Normally acid or alkaline or metal salts are required as a catalyst to facilitate the hydrolysis process.

Aluminum is a reactive metal that produces hydrogen and aluminum hydroxide when reacted with aqueous or concentrated alkaline salts. The reaction is highly exothermic, that is, it releases heat. The readers should attempt to answer the following questions to understand the characteristics of hydrogen.

1. What are the physical properties of hydrogen?
2. Explain the classifications of hydrogen production based on sources.
3. What are the physical and chemical properties of aluminum?

5.2.2 Theoretical Framework

Aqueous metal-ion hydrolysis is a displacement chemical reaction in which Al ions act as Lewis acids. The positive charge is usually from the O—H bond in the water molecule. This increases the bonds' polarity and hence, it can break easily. At about room temperature, the reaction of aluminum metal with water to form hydrogen and aluminum hydroxide can be represented by Equation 5.1.

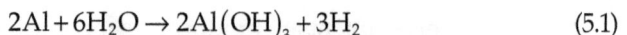

$$2Al + 6H_2O \rightarrow 2Al(OH)_3 + 3H_2 \tag{5.1}$$

The stoichiometric reaction shows that 2 moles of aluminum is required with 6 moles of water. The gravimetric and volumetric hydrogen produced in this reaction is 3.7 wt.% and 46 g/L H_2, respectively. Although this reaction is thermodynamically feasible, however, this reaction does not proceed because of the layers of aluminum oxide which cover the aluminum particles and prevent water from contacting the reactive aluminum metal [1]. Two things limit the application for hydrogen production via aluminum hydrolysis, namely: (1) an aluminum metal with pure surface must be provided to facilitate direct contact with water and (2) pure aluminum mining and production is cost-intensive and hence scraps, which require purification or surface modification, are required to promote the contact with water. Most aluminum scraps are alloys containing magnesium and silicon.

Hence, to promote this contact and sustain the reaction, the oxide layer must be removed continuously using a hydroxide catalyst or oxide promoters such as sodium hydroxide (NaOH) and aluminum oxide (Al_2O_3), and/or salts like sodium chloride (NaCl)—students can use this salt when working at home. These catalysts promote the disruption of the coherent surface aluminum oxide layer on the aluminum metal, thereby promoting the reaction with water. It must be mentioned that different promoters could form compounds of aluminum hydroxide that can be recycled. For instance, in our case, we used sodium hydroxide because it is cheap, and the reaction products are shown in Equation 5.2. However, students can use sodium chloride (table salt)

$$2Al + 6H_2O + 2NaOH \rightarrow 2NaAl(OH)_4 + 3H_2 \qquad (5.2)$$

This reaction, while not reducing the amount of hydrogen produced per mole of aluminum, introduces a new product, $NaAl(OH)_4$, which needs to be recycled or purified for use in other applications.

5.2.3 Materials

a. 2g of pure aluminum (soda and beer cans pretreated with heat to eliminate all the plastic and paint of the can).

b. 8.4 g of sodium hydroxide (NaOH).

c. Water (around 100 mL)

d. One magnetic stirrer hot plate

e. Reactor and cooler system (three-neck glass). Thermoplastics or stainless steel or other materials can be used to make the reactor, provided that it can withstand high temperature and pressure.

f. Beaker (200 mL)

g. Graduated cylinder (100 mL)

h. A balloon to collect the hydrogen.

5.2.3.1 Alternative Materials for Home Experiment

The students can also use salt promoters like sodium chloride. The procedure is as follows:

a. Aluminum powder

b. Sodium chloride

c. Deionized water

d. Blender or grinder

5.2.4 Safety Measures

5.2.4.1 General Precautions

a. **Personal Protective Equipment:** Always put on your lab coat, safety goggles, nitrile gloves, and closed-toe shoes.

b. **Proper ventilation:** Practice in a well-ventilated space or under an extractor hood.

c. **Avoid Contact:** Do not touch aluminum or solutions with bare hands.

d. **Spill Control:** In case of acid or sodium hydroxide spillage, neutralize with baking soda and clean with plenty of water.

e. **Waste Management:** Dispose of aluminum waste and solutions according to your laboratory's standards.

5.2.4.2 Specific Precautions

a. **Exothermic Reaction:** The reaction between aluminum and water is exothermic, meaning it releases heat. Use caution when handling hot materials.

b. **Flammable Hydrogen:** Hydrogen is a highly flammable gas. Do not expose open flame or sparks to the work area.

c. **Pressure in the Vessel:** The production of hydrogen can generate pressure in the vessel used. Make sure the vessel is adequate to contain the pressure and has an exhaust valve.

d. **Use of Sodium Acid or Hydroxide:** Acid and sodium hydroxide are corrosive substances. Handle with care and avoid contact with skin and eyes.

e. **Supervision:** Perform the practice under the supervision of a teacher or laboratory technician.

f. **Emergency Plan:** Know the laboratory's emergency plan in case of accidents.

g. **First Aid:** Familiarize yourself with first aid procedures for chemical burns and cuts.

It's important to remember that your safety is your first responsibility. Follow the safety measures and instructions of the teacher to avoid accidents.

5.2.5 Experimental Procedure

Figure 5.2 shows the experimental setup to produce hydrogen by hydrolysis of aluminum. The reactor is a three-neck glass reactor which will contain our reactants, i.e., aluminum and aqueous sodium hydroxide. During the hydrogen production, hot water vapor or steam and other heavy byproducts are also produced, which are trapped using two water sinks. This system allows us to obtain pure hydrogen, which is stored in the balloon. The reaction is highly exothermic and DO NOT require heating. The thermometer in the reactor is used to monitor the temperature of the system, and the other thermometer is used to monitor the water temperature (record the reading in Table 5.1). Theoretical estimation reveals that 1 g of pure aluminum produces 0.1112 g of hydrogen. This amount of hydrogen at standard conditions (1 atm and 25°C) will have a volume of 1236.92 cm^3 or 1.237 L. The amount of aluminum was used as a guide to calculate how much hydrogen can be produced because it is a limiting reactant.

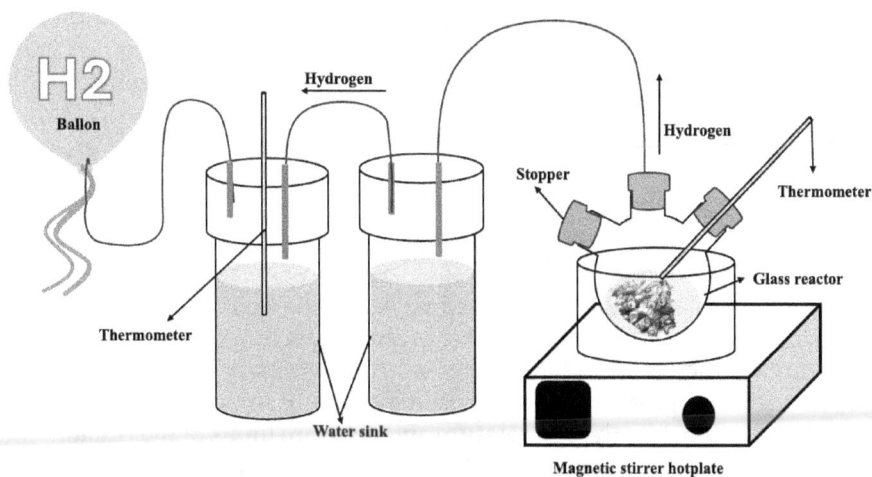

FIGURE 5.2
Experimental setup of the hydrogen production by hydrolysis.

TABLE 5.1

Temperature and Time Reading from the Reactor

Time [minutes]	Temperature [°C]

1. Weigh 8.4 g of sodium hydroxide and add 100 mL of water. Stir the solution until the NaOH pellets are completely dissolved.
2. Weigh 2 g of aluminum scraps and pour them into the reactor.
3. Pouring the aluminum into the reactor is shown in Figure 5.2.
4. Add 21 mL of the mix of sodium hydroxide dissolved in water into the reactor.
5. Close the reactor, take notes of the elapsed time and the temperature inside the reactor. The temperature can be observed using the thermometer.
6. Connect a balloon at the end of the cooling system and wait for it to inflate.

5.2.5.1 Experiment at Home

Mill the aluminum and sodium chloride salt in the ratio of 1:1–1:3 (aluminum:salt). You can use a blender or a manual grinder. You can do this for different times from 2 to 10 hours. After milling, wash off the salt using cold tap water. Then disperse the aluminum in water at 55°C and monitor the hydrogen evolution reaction. Collect the hydrogen using a gas bag. The milling of the aluminum helps to disrupt the oxide layers and hence, promotes the reaction kinetics when the aluminum is exposed to water.

5.2.6 Results

The students/readers are encouraged to report the findings of their experiment and try to use different amounts of aluminum to see what happens and how quickly they can produce hydrogen. Also, the students should monitor the temperature profile and graph the temperature versus time profile. It is important to consider the temperature and pressure limits of the reactor. The students should answer questions like: What is the mass of aluminum consumed in their experiment? What factors might influence the reaction? At what point was the highest gas production observed?

5.2.7 Conclusions

With the correct waste treatment, this method of obtaining hydrogen can be used as a way to solve the problem of aluminum scraps, but is also energy inefficient, because all the heat is lost during the reaction. However, this heat loss could be channeled to heating of processing equipment. Also, this technology can be used as a backup of the main hydrogen source in a hydrogen economy, considering the huge aluminum wastes generated globally.

5.3 Hydrogen Production by Water Electrolysis

5.3.1 Preliminary Questionnaire and Summary

Electrolysis can be generally defined in chemistry as passing direct current in a solution of liquid to realize chemical decomposition or drive a nonspontaneous reaction. In this concept, an electrochemical cell comprising two electrodes (cathode and anode electrodes), an electrolyte, and external voltage are required. The equipment used to carry out electrolysis is called an electrolyzer. The electrolyzer uses current to split water into hydrogen and oxygen. The current can also be generated from renewables such as solar, or wind. This technology was first developed in 1800 by Nicholson and Carlisle and has evolved rapidly into matured technology. Today, different types of electrolyzers exist and are mainly named based on the operating temperature and nature of the applied electrolyte. The students are encouraged to answer these questions and read further about electrolysis.

1. In water electrolysis, name the gases collected at the anode and cathode, and how do they compare?
2. Mention three electrode materials used in the electrolysis of water.
3. What is the general and half-cell chemical reaction that occurs in electrolysis at pH=14?

5.3.2 Theoretical Framework

Water electrolysis is a chemical process that involves the use of electric current to split water into its different constituents. An electric current is passed through an electrolyte solution, consisting of positively or negatively charged ions. These ions move toward the electrodes where they are reduced or oxidized. For instance, the positive ions from the electrolyte are attracted to the negative electrode, known as the cathode. At the cathode, these ions gain electrons and reduce, forming a neutral species; the same happens for the anode. Hence, electrolysis is a redox reaction. It is widely used in industry for production of chemicals, such as chlorine, hydrogen, and aluminum.

Electrolysis is carried out in an electrolyzer, which is made of two electrodes (cathode and anode), and a membrane in some cases. There are three main types of electrolyzers namely proton exchange membrane electrolyzer, alkaline electrolyzer, and solid oxide electrolyzer. These electrolyzers are named based on the type of electrolyte and temperature. These electrolyzers work differently depending on the electrolyte material. Also, temperature requirement and efficiency vary. You are encouraged to fill Table 5.2 and understand the characteristics of the different electrolyzers. For the purpose of this experiment, the proton exchange membrane (PEM) electrolyzer was used and will be briefly discussed.

TABLE 5.2

Complete the Properties of the Electrolyzers

Properties	PEM Electrolyzer	Alkaline Electrolyzer	Solid Oxide Electrolyzer	Reference
Electrolyte				
Electrode/catalyst				
Operational temperature (°C)				
Efficiency (kWh to produce 1 kg of H_2)				
Anode reaction				
Cathode reaction				

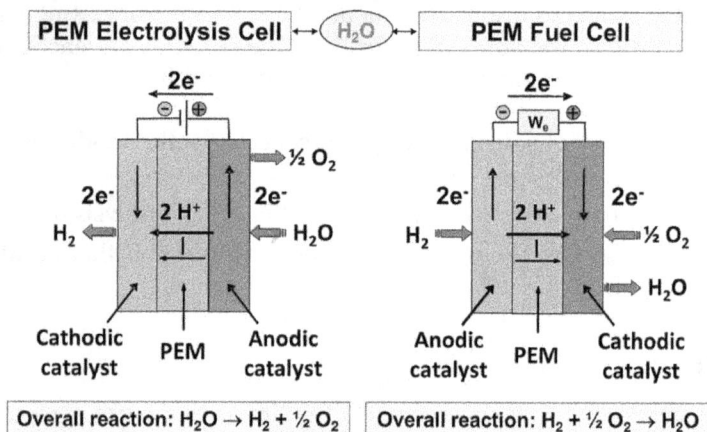

FIGURE 5.3

Schematics of PEM electrolyzer and PEM fuel cells with flow of protons. (Reproduced from Lamy [3] with copyright permission.)

The PEM electrolyzer splits water into oxygen and proton. The oxygen is released, and the proton (H^+) produced in the anode passes through a membrane (e.g., nafion membrane or proton-conducting membrane) toward the cathode [2,3]. Also, the electrons move through the external circuit. At the cathode, the electrons reduce the protons to form hydrogen gas. This electrolysis process can be reversed to have a PEM fuel cell. In the case of PEM fuel cell, the hydrogen and oxygen gases combine to produce a direct electric current with water and heat as byproduct. Figure 5.3 shows the PEM electrolyzer and fuel cell.

The fundamental configuration of a PEM electrolyzer comprises two half cells housing a slim, proton-conducting and electron-insulating Proton Exchange Membrane (PEM) positioned at the core [2]. A porous catalyst layer on either side of the membrane facilitates the necessary reactions. Together, the PEM and the dual catalyst layer constitute the Membrane Electrode Assembly (MEA). Surrounding the MEA is the current collector, establishing both physical and electrical connections between the catalyst layer and the bipolar plate. Nonetheless, it is the bipolar plate that furnishes structural robustness to the cell, creates conduits for reactants and products, and segregates one cell from another within a stack. The students are encouraged to read more on electrolyzers, design, modeling, and commercial challenges encountered in each of these electrolyzers.

The half-cell electrolyzer reactions can be represented by Equations 5.3 and 5.4.

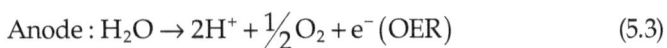

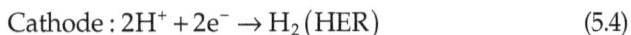

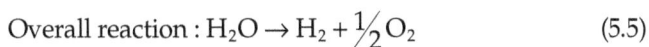

$$\text{Anode}: H_2O \rightarrow 2H^+ + \tfrac{1}{2}O_2 + e^- \text{(OER)} \tag{5.3}$$

$$\text{Cathode}: 2H^+ + 2e^- \rightarrow H_2 \text{(HER)} \tag{5.4}$$

$$\text{Overall reaction}: H_2O \rightarrow H_2 + \tfrac{1}{2}O_2 \tag{5.5}$$

where HER means hydrogen evolution reaction, and OER means oxygen evolution reaction. Equation 5.5 is the sum of the two half-cell reactions occurring within the electrolyzer.

5.3.3 Instruments, Equipment, and Materials

    a. Two 50 cm tubes filled with water

    b. Electrolyzer

    c. PEM fuel cell

    d. PEM electrolyzer (with stainless steel electrodes)

    e. Solar panel (capacity 2 V or more)

    f. Electric motor with blades (1.5 V)

    g. Alligator connection cables

    h. 500 mL of water (water density 1 g/mL) [18.0146 g/mol]

    i. Sodium hydroxide (NaOH) salt

5.3.4 Safety Measures

5.3.4.1 *General Precautions*

    a. The general safety rules mentioned in Section 3 apply to this section and should be adhered to.

    b. Also, ensure to clean any spill of the electrolyte.

5.3.4.2 Specific Precautions

a. **Electrical Hazard:** *The* practice involves the use of an electrical power source. Make sure cables and connections are in good condition. Especially the electrode connector should not be oxidized with rusts.

b. **Flammable Hydrogen:** Hydrogen is a highly flammable gas. Do not expose open flame or sparks to the work area.

c. **Explosions:** The mixture of hydrogen and oxygen can be explosive. Do not allow gas to build up in the work area.

d. **Pressure Buildup:** always ensure there is no pressure buildup in the system.

e. **Temperature:** since there is a temperature buildup during experiment, it is expected that the integrity of the plastic bottles will decrease, or they can deform.

Note: It's important to remember that safety in the lab is everyone's responsibility. Follow the safety measures and instructions of the teacher to avoid accidents.

5.3.5 Experimental Procedure

1. The experimental setup is shown in Figure 5.4. First, dissolve the sodium hydroxide salts in water.

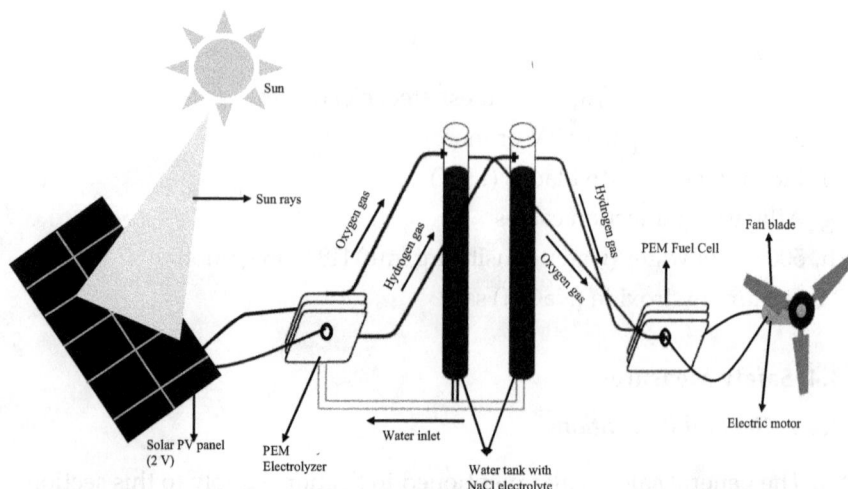

FIGURE 5.4
Hydrogen production by PEM electrolysis and energy generation using PEM fuel cell.

2. Add the dissolved sodium hydroxide in the tube and fill the tubes with water until 47 cm and close the tubes.

3. Connect up the single-cell electrolyzer with the power supply from the photovoltaic cell.

4. Ensure that the PV solar panel is tilted at 45° to expose the PV to solar radiation. In the case there is no sun, or you experience colder weather, this experiment can be conducted using any power source.

5. If the PEM electrolyzer or PEM fuel cell is not available, you can use plastics with inserted electrodes as will be shown in the next section.

6. Observe the production of the gas bubbles leaving the electrolyzer into the tube and further to the PEM fuel cell where the gases are combined.

7. Connect the motor with the small fan at the end of the PEM fuel cell to observe the rotation of the fan.

8. Analyze the experiment by measuring the voltage or current at the PEM electrolyzer as the input and the output of the PEM fuel cell. Notice the drop in voltage or current and discuss the observations.

9. In the second experiment (without video), a glass or acrylic rectangular container with plastic or Teflon cover is used (Figure 5.5).

10. Drill a round hole (½ inches) on the cap of the container and the cap of the bottles to pass the electrodes.

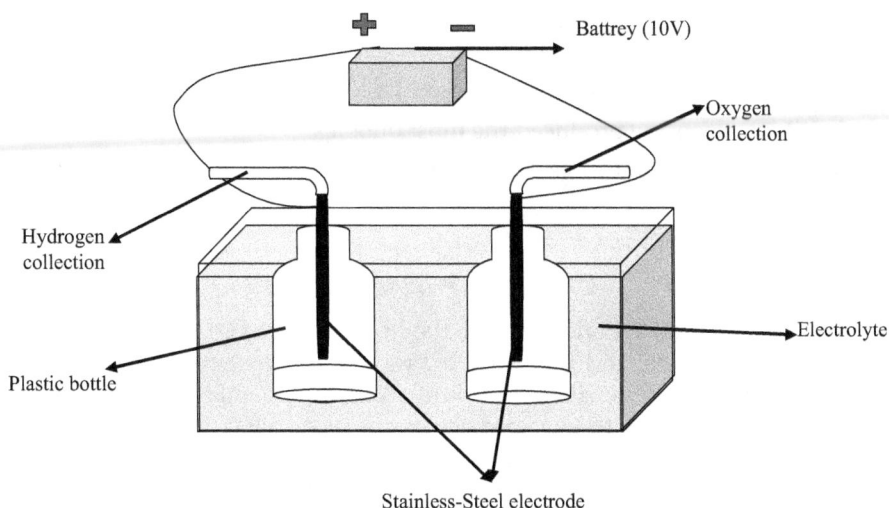

FIGURE 5.5
Hydrogen and oxygen generator using electrolytic cell.

11. Fix a small hollow ½ inches stainless steel tube on the container cap and the bottle cap. This must be fixed to align with the holes of the container and the bottle cap so that the electrodes can pass.

12. Use adhesives to secure the cap of the water bottles and ensure there is no gas leak.

13. With a knife, cut the water bottle from the bottom; it is recommended to cut it about ¼ down the water bottle.

14. Insert the electrode through the top of the stainless-steel hollow tube and ensure that your electrode has sufficient contact with the tube. Ensure that the electrode is not touching the bottom of the container and that it is sufficiently raised to prevent bubble escape from the bottom.

15. Connect a plastic flexible pipe to the stainless-steel tube to collect the produced hydrogen and oxygen gases.

16. Add water to fill the container and add sodium hydroxide inside water, at least to get a pH of 14.

17. Using a cable, connect the stainless tube with a power source (battery of at least 10 V or solar PV panel).

18. Observe the generation of gases and put an inflatable balloon to collect the hydrogen.

5.3.6 Results

Write a report about the experiments and compare the results from the two experimental setups.

Answer the following questions inside this notebook.

1. What will happen if the NaOH is not added to the water?
2. Which electrode produced the most bubbles?
3. Why is the volume of gas collected in one electrode double that of the other electrode?

5.3.7 Conclusions

In the PEM electrolyzer and fuel cell, the heat produced could be channeled to other applications, and the water byproduct can be recycled to continue the electrolysis process, since it is pure water. The combination of hydrogen and oxygen gases to generate electricity using PEM fuel cell stacks is a mature technology; however, care must be taken because this could be a potential source of intense explosion. It is recommended that the electrolyzers should be designed to have hybrid power sources in case solar PV cells cannot be used alone.

5.4 Biohydrogen Production by Dark Fermentation (DF)

5.4.1 Preliminary Questionnaire

1. What is dark fermentation and the two notable DF pathways?
2. List the factors that influence the hydrogen production rates and yield in DF and discuss the factors that can inhibit the biological hydrogen production in DF.
3. Describe briefly three processes that can integrate DF to improve viability of DF.
4. During the fermentation process to produce hydrogen, mention two byproducts of this reaction and how they can be integrated into the process.

5.4.2 Theoretical Framework

Apart from electrolysis and hydrolysis, thermochemical processes like gasification and hydrothermal processes can be used to produce hydrogen. Biological interaction of microorganisms with organic substrates is a potential viable option for biohydrogen production. Although this technology is not fully matured, however, consideration in terms of costs and energy requirements shows that biohydrogen production could significantly boost the hydrogen industry. Two broad steps have been identified, namely light-dependent and light-independent synthesis routes (Figure 5.6). The light-independent route via dark fermentation (DF) for biohydrogen is the most studied research technique because of its theoretical high production rates.

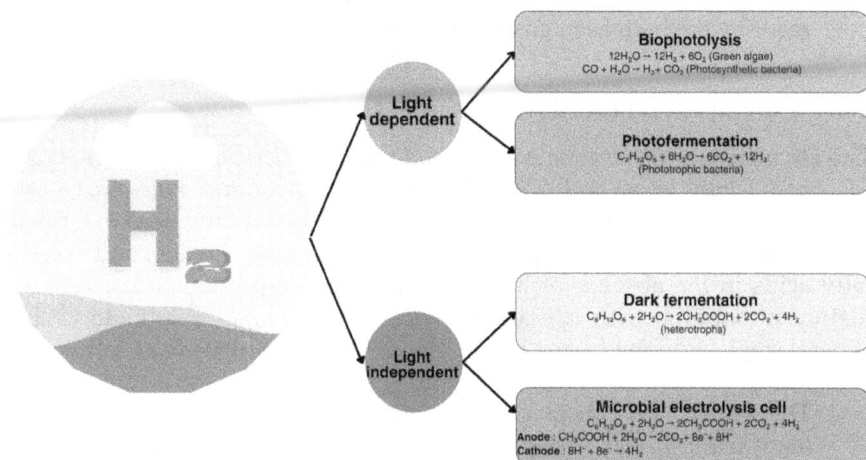

Biophotolysis
$12H_2O \rightarrow 12H_2 + 6O_2$ (Green algae)
$CO + H_2O \rightarrow H_2 + CO_2$ (Photosynthetic bacteria)

Photofermentation
$C_6H_{12}O_6 + 6H_2O \rightarrow 6CO_2 + 12H_2$
(Phototrophic bacteria)

Light dependent

Dark fermentation
$C_6H_{12}O_6 + 2H_2O \rightarrow 2CH_3COOH + 2CO_2 + 4H_2$
(heterotrophs)

Microbial electrolysis cell
$C_6H_{12}O_6 + 2H_2O \rightarrow 2CH_3COOH + 2CO_2 + 4H_2$
Anode: $CH_3COOH + 2H_2O \rightarrow 2CO_2 + 8e^- + 8H^+$
Cathode: $8H^+ + 8e^- \rightarrow 4H_2$

Light independent

FIGURE 5.6
Biohydrogen production by light-dependent and -independent pathways.

Unlike anaerobic digestion to produce biogas, DF is employed to produce biohydrogen. The system is the same, however, certain measures have to be taken to prevent anaerobic digestion. These measures include, lower hydraulic retention time (HRT), operation in pH less than 7, and pretreatment of inoculum to kill the growth of methanogens (hydrogen-consuming bacteria). The DF could be used to treat organic wastes resulting in a sustainable circular bioeconomy. The organic substrates could be sourced from carbohydrate-rich materials such as first- and second-order generation biomasses, industrial wastewater, and sludge. The pH, temperature, inoculum-to-substrate ratio, feeding rate, and the partial pressure of hydrogen are factors that affect the DF process. The technology readiness of DF is still not sufficient for industrial scaling and would require an integration with anerobic digestion to increase energy recovery and boost the viability of DF process. Also, inoculum enrichment methods and pretreatment of substrates to improve biodegradability are some of the adopted methods to improve hydrogen yield.

5.4.3 Instruments, Equipment, and Materials

a. A Gerber (usually for kids food) flask 1,000 mL with cover. If you can afford it, a jacketed reactor with several openings for feeding and removal of the gases is an ideal reactor for laboratory sessions.

b. Stirring rod and heated water bath to maintain the temperature at 35°C ± 1°C.

c. Substrate, in this case, a sludge collected from a wastewater plant.

d. Inoculum obtained from sewage sludge.

e. 50 mL measuring cylinder, inverted in a water bath for water displacement by the hydrogen gas and calculation of the yield (Figure 5.7).

5.4.4 Experimental Procedure

The DF microbes *Escherichia coli*, *Clostridium spp.* (*Clostridium acetobutylicum*, *Clostridium butyricum*, and *Clostridium pasteurianum*), and *Enterobacter spp.* (*Enterobacter cloacae* and *Enterobacter aerogenes*) are commonly used to ferment organic substrates converting it into hydrogen, carbon dioxide, and volatile fatty acids in the absence of light. These hydrogen-producing bacteria are mainly mesophilic with operational temperature range of 20°C–45°C. The general steps for biohydrogen production using the sludge are as follows:

1. The sludge was collected from sedimentation of wastewater treatment plant. It was first heated to inhibit methanogens, and the pH was adjusted to 6.5. This process is conducted to ensure there is no inhibition to the fermentation process.

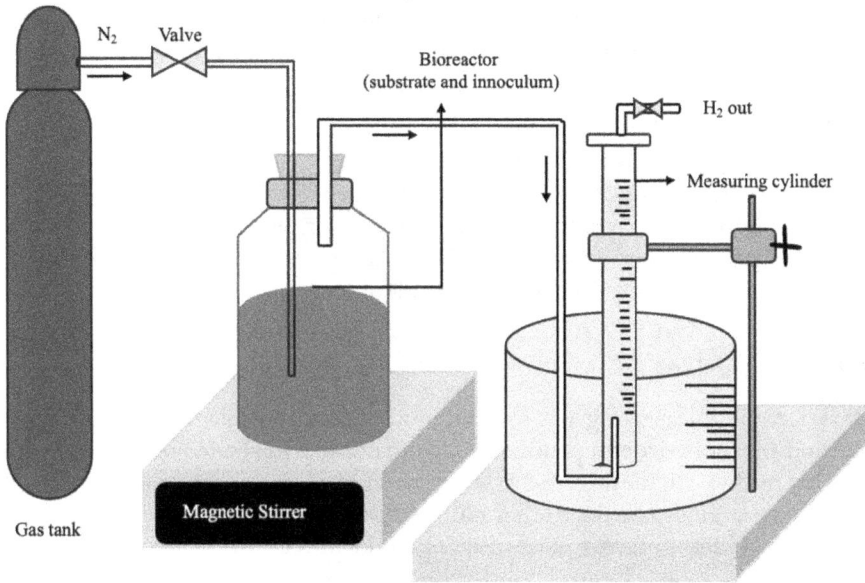

FIGURE 5.7
Experimental setup for dark fermentation process for hydrogen production using anerobic bioreactor.

2. The inoculum from sewage sludge was subjected to thermal shock at 90°C for 30 minutes to kill the methanogenic (H_2-consuming bacteria) bacteria such as homoacetogenic bacteria and methanogenic archaea bacteria.

3. The wastewater sludge was mixed with an expired orange juice as an organic carbon source.

4. This mixture with the inoculum is transferred into the fermentation vessel having a working volume of 500 mL (i.e., bioreactor) to maintain anaerobic conditions for fermentation.

5. Students should vary the temperature (35°C–40°C), substrate concentration (4–10 g-VS/L), inoculum-to-substrate ratio (0.25–1) to perform parametric optimization of the DF.

6. Adjust the solution pH to 7 using 1 M HCl and 1 M NaOH to ensure stability for the microorganisms to maintain high fermentative ability.

7. Purge the reactor with high-purity nitrogen (200 mL/min) for 5 minutes and then seal it to begin fermentation.

8. The temperature of the reactor can be controlled by inserting it in a water bath or using heating mantle or heating resistance connected with a thermometer. Also, this process can be made green by

circulating the reactor with solar-heated water. The reactor should be stirred manually with a sterilized rod for 1 minute, twice a day. Constant stirring can stress the bacteria.

9. Conduct all the experiments in duplicates over a time period of 96 hours.

10. After fermentation, students could purify the gas by passing it through gas chromatography or gas membrane to collect hydrogen. The CO_2 gas can be sent to facilitate the growth of microalgae. The sludge can be applied as a biostimulant (organic fertilizer) for crops.

5.4.5 Conclusions

Dark fermentation using wastewater sludge can be utilized as an effective method for biohydrogen production, contributing to renewable energy production and waste management. Using solar energy to heat water circulated in a jacketed glass reactor to maintain temperature will significantly reduce energy cost. Temperatures above 38°C could impact the hydrogen production rate.

References

1. C.Y. Ho and C.H. Huang, "Enhancement of hydrogen generation using waste aluminum cans hydrolysis in low alkaline de-ionized water," *International Journal of Hydrogen Energy*, vol. 41, pp. 3741–3747, 2016, https://doi.org/10.1016/j.ijhydene.2015.11.083.

2. D.S. Falcão and A.M.F.R. Pinto, "A review on PEM electrolyzer modelling: Guidelines for beginners," *Journal of Cleaner Production*, vol. 261, 2020, https://doi.org/10.1016/j.jclepro.2020.121184.

3. C. Lamy, "From hydrogen production by water electrolysis to its utilization in a PEM fuel cell or in a SO fuel cell: Some considerations on the energy efficiencies," *International Journal of Hydrogen Energy*, vol. 41, pp. 15415–15425, 2016, https://doi.org/10.1016/j.ijhydene.2016.04.173.

6

Ideal and Real Gas Home Exploration

Summary

Understanding the behavior of gases is essential in both academic and real-world applications, from engineering systems to everyday phenomena like inflating tires or using aerosol sprays. While ideal gases follow predictable laws, real gases can deviate under certain conditions. This experiment is important because it provides a simple, visual way for students to connect theory to practice. By observing how gas volume changes with temperature, learners can better grasp the principles of the Ideal Gas Law and appreciate the conditions under which real gases might behave differently. In this home experiment, students use a balloon and a household freezer to investigate the Ideal Gas Law. After inflating a balloon at room temperature, it is placed in the freezer to observe how cooling affects the gas inside. As the temperature decreases, the gas molecules slow down and the balloon visibly contracts, illustrating the direct relationship between temperature and volume described by Charles's Law. This hands-on activity offers a clear demonstration of molecular behavior in gases, reinforcing theoretical concepts through practical observation.

6.1 Introduction

Gases play a crucial role in chemical and industrial processes, making their study essential for fields such as chemistry, physics, and engineering. Understanding how gases behave under different conditions allows scientists and engineers to predict their interactions in various applications, from manufacturing and energy production to environmental monitoring and medical technology. The study of gas laws provides fundamental insights into how gases respond to changes in pressure, volume, and temperature, helping to optimize processes such as combustion, refrigeration, and chemical synthesis. These principles are not only theoretical but also have direct implications for real-world problem-solving in scientific and industrial settings.

DOI: 10.1201/9781003374503-6

The fundamental gas laws including the Boyle's Law, Charles's Law, Avogadro's Law, and the Ideal Gas Law helps to describe the relationships between pressure, volume, temperature, and the amount of gas present in a system. Boyle's Law states that for a fixed amount of gas at constant temperature, pressure and volume are inversely proportional. Charles's Law establishes that at constant pressure, the volume of a gas is directly proportional to its absolute temperature. Avogadro's Law explains that equal volumes of gases at the same temperature and pressure contain an equal number of molecules. Combining these principles leads to the Ideal Gas Law, expressed as $PV = nRT$, which serves as a foundational equation in gas behavior analysis. These laws assume ideal conditions, where gas molecules do not interact and occupy negligible volume, but real gases often deviate from these assumptions under high pressures or low temperatures.

6.2 Understanding Gases

Gases are one of the fundamental states of matter, alongside solids, liquids, and plasmas. Unlike solids and liquids, gases do not have a fixed shape or volume; instead, they expand to completely fill their container, taking on its shape. This ability to expand makes them highly compressible, and they are significantly less dense than solids and liquids. Because gas particles move freely and at high speeds, gases exhibit fluid-like behavior, meaning they can flow just like liquids do. One of the most important characteristics of gases is their ability to mix completely with other gases, a process known as diffusion. Additionally, gases can pass through small openings without collisions, which is referred to as effusion [1].

To describe and quantify gases, scientists use several measurable properties. Pressure is a key property, defined as the force exerted by gas particles when they collide with the walls of their container. It is commonly measured in units such as atmospheres (atm), pascals (Pa), or millimeters of mercury (mmHg). The volume of a gas, which refers to the amount of space it occupies, is typically measured in liters (L) or cubic meters (m³). Another important factor is temperature, which is a measure of the average kinetic energy of gas particles. In gas law calculations, temperature must always be expressed in Kelvin (K) to ensure consistency. Lastly, the amount of gas present is quantified in terms of moles (n), which provides a count of the number of gas molecules in a given sample.

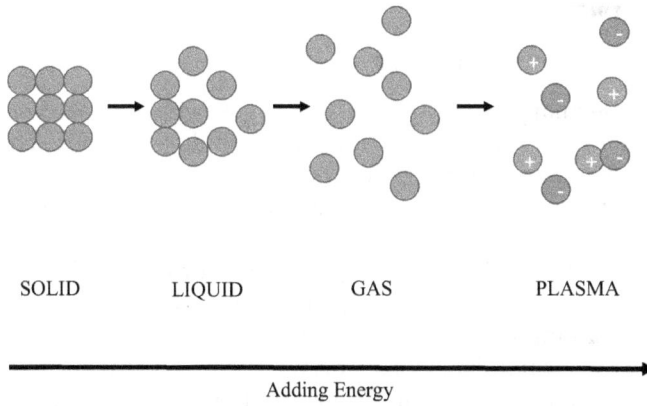

SOLID   LIQUID   GAS   PLASMA

Adding Energy

1. What makes a gas different from a solid, liquid, or plasma?

_____

_____

_____

_____

_____

2. Describe a real-world example of gas diffusion or effusion and explain how it relates to the properties of gases discussed above.

_____

_____

_____

_____

_____

_____

3. Which of the following statements about gases is TRUE?
   a. Gases have a fixed shape but not a fixed volume.
   b. Gas particles move freely and at high speeds.
   c. Gases are incompressible.
   d. The pressure of a gas is not affected by temperature.
   (Answer: B)

6.3 Kinetic Molecular Theory

The kinetic molecular theory (KMT) explains gas behavior based on the idea that gas particles are in constant motion and collide frequently. This theory is built upon five main assumptions. First, gas particles move randomly and travel in straight lines until they collide with another particle or the walls of their container. Second, when these collisions occur, they are perfectly elastic, meaning that no kinetic energy is lost. Third, the volume occupied by individual gas molecules is negligible compared to the total volume of the gas itself, making gases largely empty space. Fourth, there are no significant attractive or repulsive forces between gas molecules, meaning that their movement is influenced only by external factors such as pressure and temperature. Lastly, the average kinetic energy of gas particles is directly proportional to the temperature of the gas in Kelvin. This means that as temperature increases, gas particles move more quickly.

Gas particles are in constant, random motion.

Gas particles are tiny compared to the distances between them.

Collisions between gas particles (and with the container) are perfectly elastic.

There are no intermolecular forces between gas particles.

The average kinetic energy of gas particles is directly proportional to temperature in Kelvin.

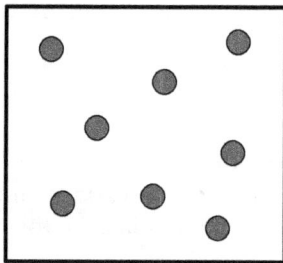

4. If the average kinetic energy of a gas particle is directly proportional to temperature, what is the ratio of the average kinetic energy of gas molecules at 400 K compared to 200 K?

_____

_____

_____

_____

(Answer: Since kinetic energy is proportional to temperature, the ratio is 400/200 = 2. The kinetic energy at 400 K is twice that at 200 K.)

The kinetic molecular theory serves as the foundation for understanding gas behavior and helps explain why gases respond predictably to changes in pressure, volume, and temperature. While real gases do not always follow these assumptions perfectly, the theory provides a useful approximation in many cases and forms the basis for the ideal gas law.

6.4 What Is an Ideal Gas?

An ideal gas is a theoretical concept that represents a gas whose behavior perfectly follows a set of assumptions that simplify its interactions. These assumptions allow scientists and engineers to predict gas behavior under different conditions using mathematical equations. While no real gas behaves ideally under all conditions, the model provides a close approximation that is accurate for many practical applications. The assumptions of an ideal gas include:

- Gas molecules are constantly moving in random directions
- Gas molecules' collisions with the walls of their container are completely elastic.
- Each molecule acts as a perfectly spherical point mass, meaning its physical mass takes up no significant volume.
- No intermolecular forces, meaning that individual gas molecules neither attract nor repel one another.

The **Ideal Gas Law** is an ensemble of four fundamental gas laws that describe the relationships between pressure, volume, temperature, and the number of gas molecules in a system. These laws, developed through experimental observations, help explain how gases behave under various conditions. While each law describes a specific aspect of gas behavior, their combined principles form the foundation for the Ideal Gas Law, a powerful equation used in scientific and industrial applications.

Boyle's Law states that for a fixed amount of gas at a constant temperature, pressure and volume are inversely related. This means that as pressure increases, volume decreases, and vice versa. An everyday example of this phenomenon is a syringe: when the plunger is pushed down, the volume of trapped air inside decreases, causing the pressure to rise. Similarly, divers experience the effects of Boyle's Law when they ascend or descend in water. As they go deeper, the increased water pressure compresses the air in their lungs, requiring them to adjust their breathing to prevent lung overexpansion when resurfacing. Boyle's Law is mathematically expressed as:

$$\text{Boyle's Law: } P_1V_1 = P_2V_2$$

Charles' Law describes the direct relationship between volume and temperature when pressure remains constant. According to this law, increasing the temperature of a gas causes its volume to expand, while cooling a gas causes it to contract. This principle is evident in a balloon: when exposed to heat, the gas molecules inside move more rapidly and spread out, making the balloon expand. Conversely, if the balloon is placed in a freezer, the gas molecules slow down, causing the balloon to shrink. Charles' Law plays a crucial role in hot air balloon operation, where heating the air inside the balloon increases its volume and makes it less dense than the surrounding air, allowing it to rise. The mathematical expression for Charles' Law is:

$$\text{Charles' Law: } V_1/T_1 = V_2/T_2$$

Gay-Lussac's Law states that at constant volume, the pressure of a gas is directly proportional to its temperature. This means that when the temperature of a gas increases, its pressure also increases, provided that the volume remains unchanged. A practical example of this law is the danger posed by aerosol cans in extreme heat. If a sealed can is left in direct sunlight or near a heat source, the increasing temperature causes the gas inside to expand and exert more pressure on the walls of the container. If the pressure exceeds the can's structural limits, it may rupture or explode. This law is also important in internal combustion engines, where the rapid heating of fuel-air mixtures in a fixed-volume cylinder leads to high pressure, driving the movement of pistons. Gay-Lussac's Law is mathematically represented as:

$$\text{Gay-Lussac's Law: } P_1/T_1 = P_2/T_2$$

Finally, Avogadro's Law establishes that at constant temperature and pressure, the volume of a gas is directly proportional to the number of gas molecules present. This means that adding more gas to a container increases its volume, assuming that the pressure and temperature remain the same. A common example is inflating a balloon: as more air is pumped in, the balloon expands because the increased number of gas molecules pushes outward against the balloon's elastic walls. This law is particularly useful in stoichiometric calculations in chemistry, where it helps determine the volumes of gases involved in chemical reactions. Avogadro's Law is expressed as:

$$\text{Avogadro's Law: } V_1/n_1 = V_2/n_2$$

By combining these four gas laws, scientists derived the ideal gas law, which allows for the calculation of any one variable when the others are known.

$$\text{Ideal Gas Law: } PV = nRT$$

5. A gas occupies 5.0 L at a pressure of 2.0 atm. If the pressure is increased to 5.0 atm while the temperature remains constant, what will be the new volume of the gas?

_____

_____

_____

_____

(Answer: Using Boyle's Law, $V_2 = (P_1 V_1)/P_2 = (2.0\ atm \times 5.0\ L)/5.0\ atm = 2.0\ L)$

6. If 2 moles of gas occupies 4.0 L at a pressure of 2.3 atm, what temperature will the gas be held at?

_____

_____

_____

_____

(Answer: Using the ideal gas law, $T = PV/nR = (2.3\ atm \times 4.0\ L)/(2\ mol \times 0.0821\ Latm/molK)) = 56\ K)$

7. If the temperature of a gas doubles (in Kelvin) while the pressure remains constant, what happens to its volume?
   a. It stays the same.
   b. It doubles.
   c. It halves.
   d. It increases but not necessarily doubles.
   (Answer: B)

6.5 Home Experiment: Investigating the Ideal Gas Law with a Balloon and a Freezer

To explore the relationship between temperature and volume in a gas system, test the predictions of the Ideal Gas Law, and observe deviations that indicate real gas behavior [2].

6.5.1 Materials Needed

- A balloon
- A measuring tape or ruler
- Baking soda and vinegar
- A freezer/refrigerator (this will work better in a freezer though!)
- A thermometer (optional, for measuring temperature)
- A notebook for recording observations
- A scale
- An empty water bottle (or other plastic bottle, the size doesn't matter)

6.5.2 Safety Measures

1. **Wear Safety Goggles and Gloves**: Always wear safety goggles to protect your eyes from any potential splashes of vinegar or baking soda. Gloves should be worn to avoid skin irritation or contact with the chemicals.

2. **Use the Balloon Properly**: Ensure the balloon is securely attached to the water bottle to prevent it from popping off unexpectedly during inflation. Use caution when handling the balloon after it is inflated to avoid any pressure buildup.

3. **Handle Vinegar and Baking Soda Carefully**: Both vinegar and baking soda are generally safe, but avoid direct contact with eyes or skin for prolonged periods. If contact occurs, rinse thoroughly with water.

4. **Freezer Safety**: Be cautious when handling the balloon after removing it from the freezer, as it will be very cold. Avoid direct skin contact with the cold surface of the balloon to prevent frostbite. Do not place the balloon in the freezer for too long, as it could potentially burst if the temperature drops too low, or it may cause discomfort when handling it.

5. **Use Appropriate Containers**: Ensure that the water bottle you are using is sturdy enough to handle the reaction. Avoid using glass containers, which may break under pressure.

6. **Dispose of Materials Properly**: After the experiment, dispose of the vinegar and baking soda mixture by diluting it with water and pouring it down the drain. Clean up any residue promptly.

6.5.3 Procedure

1. **Inflate the Balloon:** Add 5 g of baking soda and an excess of vinegar (add at least 10 mL but you can add more) to an empty water bottle.

Quickly attach the balloon on top and watch it inflate. Once the balloon stops filling, pull it off the water bottle and tie it shut (Figures 6.1 and 6.2).

FIGURE 6.1
Measure out 5g of baking soda using a scale.

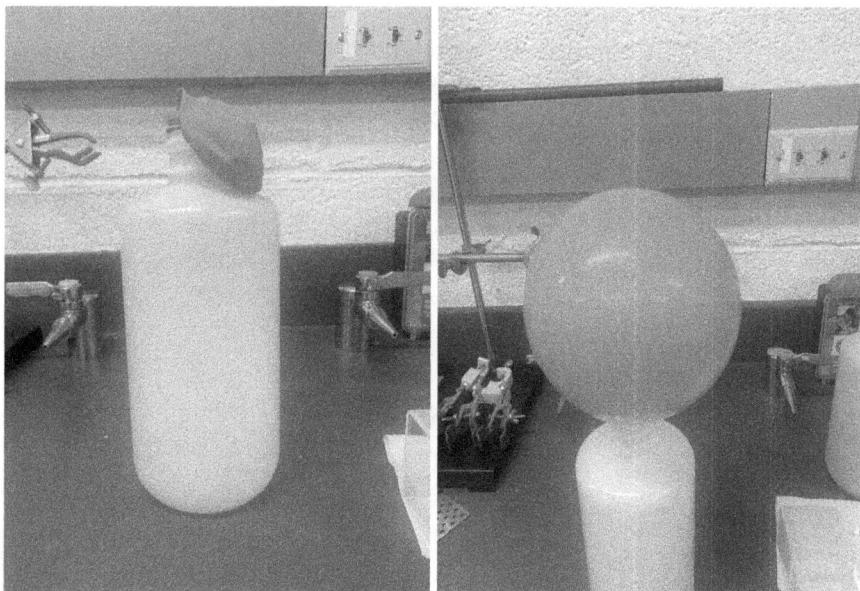

FIGURE 6.2
Add the 5g of baking soda and an excess amount of vinegar into a clear plastic water bottle. Secure the balloon firmly on top of the bottle and allow it to expand (like shown in right picture).

2. **Measure the Initial Volume:** Use a measuring tape to find the circumference of the balloon at room temperature (Figure 6.3). Convert this to an approximate volume using the formula for a sphere: $V = \dfrac{4}{3}\pi r^3$

3. **Place the Balloon in the Freezer:** Leave the balloon in the freezer for at least 30 minutes. If available, record the freezer temperature. If not, estimate that the temperature is −18°C.

4. **Remove and Measure Again:** Take the balloon out and immediately measure its circumference again. Convert this to volume using the same method.

5. **Make a Prediction:** According to the ideal gas laws, should the balloon expand or shrink? Why?

_____

_____

_____

_____

_____

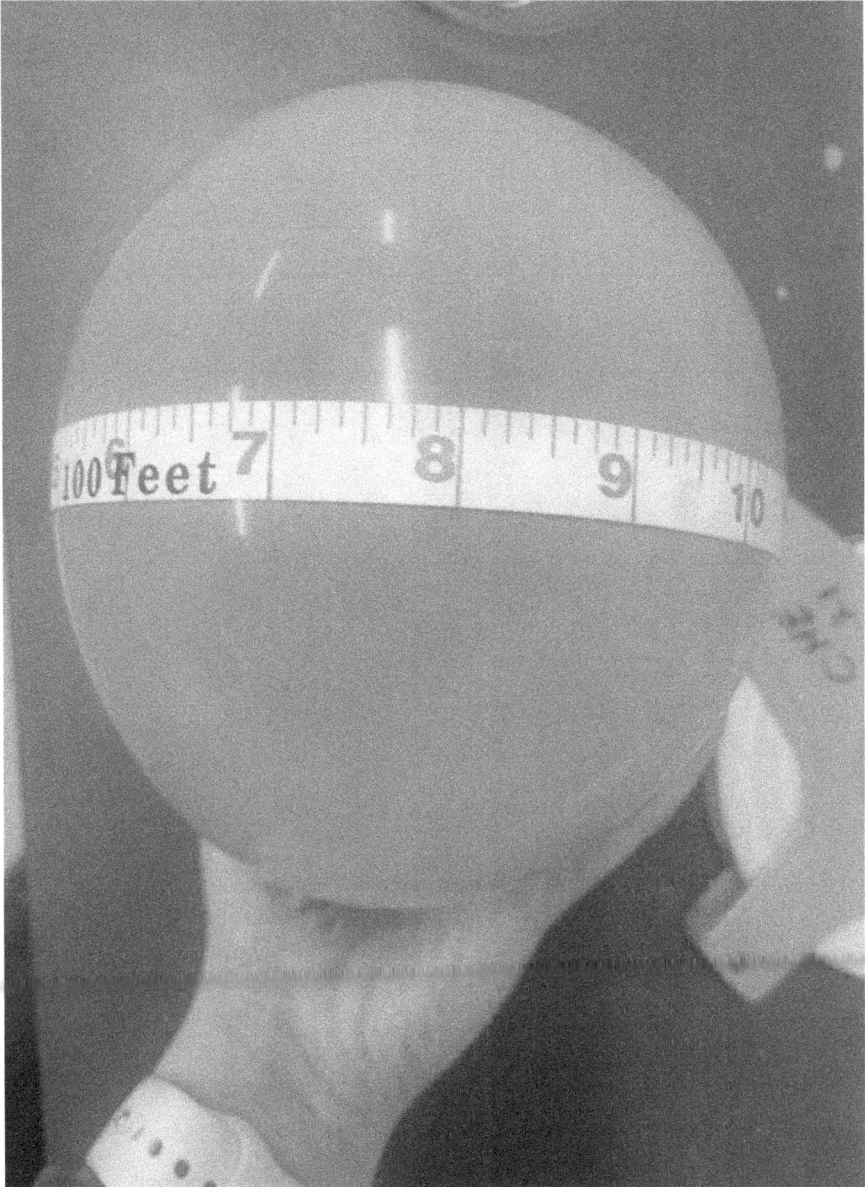

FIGURE 6.3
Use a flexible measuring tape to measure the circumference of the balloon, and then calculate
the radius from the circumference.

6. **Compare to the Ideal Gas Law Prediction:** Using the equation $V_1/T_1 = V_2/T_2$, predict what the final volume *should* be assuming ideal gas behavior.

_____

_____

_____

_____

_____

_____

7. **Observe Differences:** Compare the predicted volume to the actual measured volume.

_____

_____

_____

_____

_____

6.5.4 Reflection

1. Did the balloon shrink as expected based on the Ideal Gas Law?

_____

_____

_____

_____

_____

2. Was there a significant difference between the measured and predicted volume? Was the volume less, more, or about the same as what was predicted?

_____

_____

_____

_____

_____

_____

3. What might explain any deviations? Consider factors such as gas
   condensation, intermolecular forces, or the elasticity of the balloon
   material.

_____

_____

_____

_____

_____

_____

6.6 Real Gases and Deviations from Ideal Behavior

While the Ideal Gas Law provides a useful approximation of gas behavior,
real gases do not always adhere perfectly to its predictions. This discrepancy
arises because the assumptions made in the kinetic molecular theory such
as the lack of intermolecular forces and the negligible volume of gas molecules, do not hold under all conditions. The Ideal Gas Law assumes that gas
particles move independently of one another, with no attractions or repul
sions, and that their individual volumes are insignificant compared to the
overall volume of the container. While these assumptions work well under
low-pressure and high-temperature conditions, they break down when gases
are subjected to extreme environments.

In reality, gas molecules do experience intermolecular forces, such as van
der Waals interactions, which influence their motion and behavior. At high
pressures, gas molecules are forced closer together, making these attractive
or repulsive forces more pronounced. Attractive forces, such as London dis-
persion forces or dipole-dipole interactions, reduce the effective pressure
exerted by the gas, as some molecules pull others back toward the bulk rather
than colliding with the container walls. Conversely, at very high pressures,
repulsive forces between molecules become significant, causing the gas to
resist compression more than predicted by the Ideal Gas Law. These deviations impact processes such as gas liquefaction, where intermolecular attractions must be considered when cooling a gas to its liquid state.

Additionally, the finite size of gas molecules becomes important when the gas is compressed into a small volume. The Ideal Gas Law assumes that gas molecules occupy no space, but in reality, each molecule has a measurable volume. As pressure increases and the available space decreases, the actual volume of the gas molecules themselves becomes a larger fraction of the total volume. This effect causes the gas to take up more space than predicted by the Ideal Gas Law, leading to deviations that become especially evident in highly compressed gases, such as those stored in pressurized tanks. These deviations are accounted for in real gas equations, such as the van der Waals equation, which introduces correction factors for both intermolecular forces and molecular volume.

Understanding these deviations from ideal behavior is crucial for accurate predictions in both scientific research and industrial applications. Engineers and chemists must consider real gas behavior when designing systems that involve high pressures and low temperatures, such as refrigeration cycles, deep-sea diving gas mixtures, and the storage and transport of liquefied gases. By refining gas models with empirical corrections, scientists can develop more precise methods for analyzing and controlling gas behavior in real-world scenarios.

6.7 When Do Gases Deviate from Ideal Behavior?

Real gases tend to deviate from ideality under these primary conditions [3]:

1. **High Pressure:** As pressure increases, gas molecules are forced closer together. The assumption that their volume is negligible no longer holds because the molecules themselves occupy a measurable fraction of the total space. This results in a lower-than-expected volume compared to ideal predictions.

2. **Low Temperature:** At lower temperatures, gas molecules move more slowly, and intermolecular attractions (such as van der Waals forces) become more significant. These attractions cause molecules to cluster together rather than move freely, reducing the pressure exerted on the container walls compared to what the Ideal Gas Law predicts.

3. **High Gas Density:** Even at moderate pressures, if the number of gas molecules per unit volume is high (such as in confined spaces), intermolecular interactions become more significant, leading to deviations from ideal gas assumptions.

4. **Strong Intermolecular Forces:** Gases with strong van der Waals forces, such as hydrogen bonding in water vapor or dipole-dipole interactions in polar gases (e.g., ammonia, hydrogen chloride),

deviate more significantly from ideality than nonpolar gases like noble gases.

5. **Large Molecular Size:** The Ideal Gas Law assumes that gas molecules are point particles with negligible volume. However, gases composed of large or complex molecules (e.g., heavy hydrocarbons) occupy more space, causing deviations.

6. **Phase Changes and Near the Condensation Point:** When a gas is close to its condensation (liquefaction) temperature, intermolecular attractions dominate, leading to significant deviations from ideal behavior. For example, steam at high pressure behaves less ideally as it approaches the conditions for condensation into liquid water.

To account for these deviations, the **van der Waals equation** modifies the Ideal Gas Law by introducing correction factors for molecular volume and intermolecular forces:

$$\left(P + \frac{an^2}{V^2}\right)(V - nb) = nRT$$

where:

- a corrects for **intermolecular attractions** (lowers pressure)
- b corrects for **molecular volume** (reduces available volume)
- Other variables remain the same as in the Ideal Gas Law

The **van der Waals constants** a and b differ for each gas, meaning that some gases behave more ideally than others. For example, helium, a noble gas with weak intermolecular forces, behaves nearly ideally under a wide range of conditions, whereas water vapor, which forms hydrogen bonds, deviates significantly.

In the Ideal Gas Law, gas molecules are assumed to experience no attractive or repulsive forces. However, real gases exhibit intermolecular attractions, especially at low temperatures and high pressures. These attractions cause molecules to cluster together, leading to a lower effective pressure than what the Ideal Gas Law predicts. To account for this, the constant "a" is added to the pressure in the equation. The constant "a" quantifies the strength of intermolecular forces:

- **Larger a values** correspond to gases with strong intermolecular forces (e.g., water vapor, ammonia).
- **Smaller a values** correspond to weakly interacting gases (e.g., noble gases like helium and neon)

The Ideal Gas Law assumes that gas molecules have no volume and that the space available for gas movement is equal to the total container volume.

However, real gas molecules do occupy space, reducing the available volume for movement. The term "*b*" represents the finite volume of gas molecules, effectively subtracting a portion of the total volume to give the "free" space available. The larger the molecules, the larger the value of *b*:

- **Larger *b* values** correspond to gases with larger, bulkier molecules (e.g., hydrocarbons like butane)
- **Small *b* values** correspond with gases like helium have very small *b* values.

To further explore real gas behavior, students should **repeat the experiment** and compare their results using the van der Waals equation instead of the Ideal Gas Law.

1. **Collect Data Again:** Using the same setup as before (a balloon filled with gas and exposed to temperature changes), record new measurements of pressure, temperature, and volume.

2. **Apply the van der Waals Equation:** Instead of using the Ideal Gas Law to calculate the expected volume, now use the VDW equation with the appropriate constants: $a=3.658$ barL^2/mol^2 and $b=0.0.04286$ L/mol [4]. Assume that the baking soda is pure, so you can use the molar mass to estimate the amount of moles present.

3. **Compare the Predictions:** Determine whether the van der Waals equation provides a more accurate estimate of gas behavior, especially at lower temperatures when real gas effects are more significant. What does that mean about air at low temperatures being estimated as an ideal gas?

_____

_____

_____

_____

_____

_____

6.8 Conclusion

In this experiment, we investigated the behavior of carbon dioxide gas under changing temperature conditions by inflating a balloon using the reaction of baking soda and vinegar. The initial and final volumes of the balloon were measured and compared to predictions made using the Ideal Gas Law and the Van der Waals equation.

Our results showed that the volume of the balloon decreased when placed in the freezer, confirming that gas volume decreases with temperature, as expected from the gas laws. However, the real gas calculations using the Van der Waals equation accounted for intermolecular forces and molecular size, leading to a more accurate prediction than the Ideal Gas Law, which assumes ideal gas behavior. The difference between the two models highlights the importance of considering real gas effects, especially at lower temperatures where intermolecular forces become more significant.

This experiment demonstrates how gas laws can be applied to real-world scenarios and emphasizes the limitations of the Ideal Gas Law when dealing with non-ideal conditions. Further experiments could involve varying the amount of reactants, using different gases, or testing at different pressures to explore deviations from ideal behavior in more depth.

References

1. "10: Gases - Chemistry LibreTexts." Accessed: Apr. 08, 2025. [Online]. Available: https://chem.libretexts.org/Bookshelves/General_Chemistry/Map%3A_Chemistry_-_The_Central_Science_(Brown_et_al.)/10%3A_Gases
2. "Vinegar and Baking Soda Balloon | Activity | Education.com." Accessed: Apr. 08, 2025. [Online]. Available: https://www.education.com/activity/article/vinegar-and-baking-soda-balloon/

3. "Deviations from Ideal Gas Law Behavior." Accessed: Apr. 08, 2025. [Online]. Available: https://chemed.chem.purdue.edu/genchem/topicreview/bp/ch4/deviation5.html
4. "Non-ideal Gas – Van der Waal's Equation and Constants." Accessed: Apr. 08, 2025. [Online]. Available: https://www.engineeringtoolbox.com/non-ideal-gas-van-der-Waals-equation-constants-gas-law-d_1969.html

7

Home Production of Supercapacitors

Summary

Energy storage is crucial to sustain the transition to renewable energy. Supercapacitor is a novel energy storage device that possesses high cycle life, durability, appreciable energy density, and can be applied across a broad range of appliances. To produce supercapacitor, common materials such as aluminum foil, copper tapes, cooking salts, and coffee filters can be used at home. The capacity of stored energy when charged with a battery can light a small LED bulb and can be determined using the multimeter. This lab guide demonstrates the assembly of a supercapacitor cell using readily available materials at home. The preparation of electrode ink using commercial activated carbon with high surface area, a simple glue, and coating on the current collector are explained. Innovations in the configuration of the supercapacitor cells and electrode materials can enhance efficiency and storage of the supercapacitor.

7.1 Introduction

Supercapacitors, alternatively termed electrochemical capacitors, are energy storage devices that distinguish themselves through their energy storage mechanism. Unlike batteries, which rely on chemical reactions to store energy, supercapacitors store electrical energy via the electrostatic separation of charge. Supercapacitors serve as efficient energy storage devices, commonly employed in portable electronics and power temperature range, superior durability compared to batteries, and reliable safety performance [1]. Their advantages include extended cycle life, operation across a broad appliances, and durability. Figure 7.1 shows a supercapacitor cell, which contains a current collect, positive and negative electrodes, electrolyte, and a separator.

Supercapacitors exhibit relatively high energy density; it remains lower than that of most conventional batteries. Energy storage in supercapacitors occurs through two primary mechanisms: pseudocapacitance and electrical double-layer capacitance [2]. In the case of double-layer capacitors, ion adsorption and desorption take place at the electrode/electrolyte interface

DOI: 10.1201/9781003374503-7

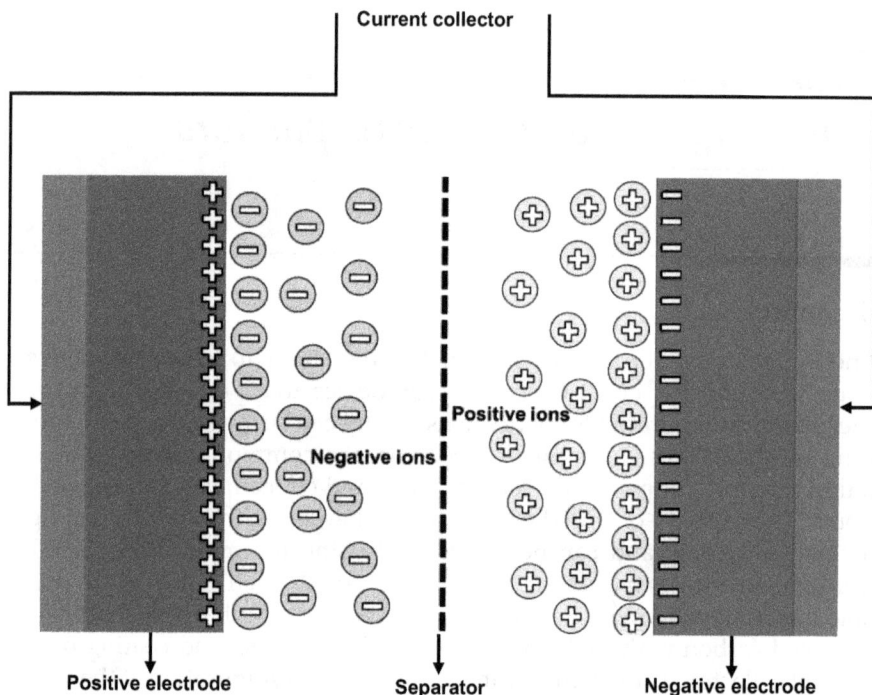

FIGURE 7.1
Supercapacitor cell assembly.

when an electric potential is applied. Various electrolytes, including acidic, alkaline, and organic types, can be utilized for this purpose. While electrical double-layer capacitors dominate the commercial market, their energy density remains a limiting factor. The energy storage and release process relies on physisorption, making the electrode's ion storage capacity dependent on its textural properties and surface functionalities. Ultimately, supercapacitor performance is governed by electrode ionic conductivity, electrolyte composition, and thermal stability.

Carbon-based electrodes are widely favored for supercapacitors due to their chemical stability, adjustable porosity, large surface area, and ease of surface modification [3]. Moreover, their production from abundant waste materials promotes sustainability. Activated carbon, in particular, stands out for its high surface area and hierarchical pore structure, predominantly micropores, which are advantageous for energy storage [4,5]. Lignocellulosic biomass has been extensively employed to produce activated carbon for supercapacitor electrodes. Activated carbon offers significant benefits compared to other electrode materials, including pure metal oxides, metal-organic frameworks (MOFs), covalent organic frameworks (COFs), and conductive polymers. Its large surface area and well-organized pore network provide ample sites for charge storage via electric double-layer capacitance, ensuring efficient ion transport and enabling rapid energy delivery with high power output. Unlike metal oxides, which rely on redox reactions and suffer structural degradation over time, activated carbon demonstrates excellent cycling stability and resilience.

The production of activated carbon can either be physical activation or chemical activation. In physical activation, two-stage temperature treatment is usually required. The first stage carbonizes the biomass at the temperature range of 500°C–800°C for 1 hour under nitrogen or argon atmosphere. In the second stage, the carbonized material is subjected to high temperature from 800°C–1,100°C employing activating agents like CO_2 and steam. One of the advantages of this process is that it does not involve chemicals that could become a potential pollution source. However, control of the pore shape and size is very difficult, and high temperature application is energy intensive. However, chemical activation makes use of chemical agents, mainly salts of alkali (potassium carbonate, sodium hydroxide, potassium hydroxide, etc.) and acids (sulfuric and phosphoric acids) to react with the lignin and other structural components of biomass. This reaction with temperature treatment reduces the energy necessary to valorize biomass into activated carbon. The chemical activation requires a temperature up to 800°C and the pores are well-defined. Also, they possess very high surface area because of their high micropore volume. The chemical agents act like a fingerprint and help in pore development. However, washing steps require significant water usage, and the effluent can be a source of pollution if not properly managed.

This comprehensive guide provides a detailed walkthrough of a safe and accessible, acid-free methodology for constructing a rudimentary supercapacitor at home, utilizing readily obtainable materials. Furthermore, it offers insights into the underlying principles and techniques employed in the fabrication of other energy storage devices, empowering enthusiasts and educators alike to delve into the captivating realm of electrochemistry.

7.2 General Objectives of the Experiments

At the end of this chapter, the students are expected to learn the following:

1. Understand the theoretical framework of supercapacitors and distinguish between a battery and a capacitor.
2. Perform experiments at home or at least at lab scale to produce material for supercapacitors.
3. Learn the basic best practices to assemble electrodes and a supercapacitor cell.
4. Use a small LED light to test the supercapacitors charged by a basic battery.

7.2.1 Answer the Following Questions about Supercapacitors

1. What are the main components or materials of a supercapacitor?

_____

_____

_____

2. Describe how you can measure capacitance with LCR meters.

_____

_____

_____

3. Describe what would be the consequences of reverse polarity of a supercapacitor cell.

_____

_____

_____

4. What are the differences between EDLC, pseudo-capacitors, and hybrid supercapacitors based on charge storage mechanism?

_____

_____

_____

7.3 Materials Required

The subsequent materials are requisite for the construction of the supercapacitor:

- **Electrode Materials:** High-surface-area materials are of paramount importance for good charge storage. Activated carbon, readily available in powdered form, presents an excellent choice. Graphite powder, frequently employed as a lubricant, can also be utilized, albeit its performance may be marginally inferior. Experimentation with alternative materials such as graphene or carbon nanotubes (assuming accessibility) may further augment performance, but these necessitate more specialized procurement. Hence, activated carbon and carbon graphite will be used.

- **Current Collectors**: These components furnish an electrical conduit to and from the electrodes. Aluminum foil, commonly found in kitchens, serves as a suitable current collector. Copper tape, procurable at electronics retailers, can additionally be employed. It is imperative to ascertain that the chosen material is highly conductive and chemically inert in the presence of the electrolyte. Aluminum foil, which is cheaper, can be used.

- **Separator**: The separator avoids direct electrical contact between the electrodes while concurrently permitting ion flow. Filter paper, routinely utilized for coffee or laboratory filtration, is a readily accessible option. Alternatives encompass porous membranes or even certain varieties of cloth, provided they are non-conductive and chemically compatible with the electrolyte.

- **Electrolyte (Safe Alternative)**: The electrolyte facilitates ion migration between the electrodes. A saltwater solution, formulated by dissolving common table salt (NaCl) in purified water, constitutes a safe and efficacious choice for this project. A baking soda ($NaHCO_3$) solution can likewise be employed. The purity of the water is very important; distilled or deionized water is preferred to minimize undesirable reactions.

- **Container**: The container is for structural support to contain the electrode assembly and avoid leakage of electrolyte. A plastic casing, such as a small box, is ideal. Any non-conductive, chemically inert material can be employed. Contemplate the dimensions of the assembled supercapacitor when selecting the container.

- **Conductive Glue or Binder**: This material guarantees optimal electrical contact between the electrode material and the current collector. A mixture of graphite powder and a standard adhesive, such as epoxy or cyanoacrylate glue, can be utilized. The glue should be applied sparingly to avert impeding ion transport. You can use a white glue, it will serve, but you need to do a thorough mixing.

- **Leads and Connectors**: These components permit external connection to the supercapacitor. Wires with alligator clips or metal tabs are suitable for this purpose. Ensure secure electrical contact between the leads and the current collectors.

7.4 Methodology

7.4.1 Electrode Preparation

- **Grinding and Pulverization**: In the event of utilizing activated carbon in bulk form, gently crush and grind it into a fine powder

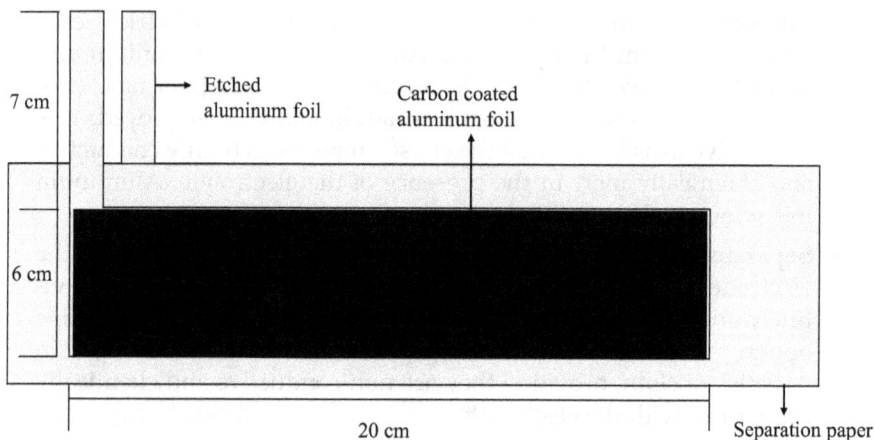

FIGURE 7.2
Supercapacitor cell assembly demonstration with dimensions.

utilizing a mortar and pestle or a comparable method. The finer the powder, the greater the surface area accessible for charge storage.

- **Preparation of Conductive Paste**: Thoroughly mix the powdered carbon (20 g) with about 2 g of white glue and 50 mL of deionized water until you have a viscous paste. The consistency should be similar to that of toothpaste. Refrain from employing excessive glue, as this can impede the flow of ions. Cut the aluminum foil rectangular of 20×6 cm with an extended point of 7 cm for connection to the voltage source or text for discharge (Figure 7.2).

- **Drying the Electrodes**: Allow the painted aluminum to dry very well, this can take several hours depending on the humidity and climate conditions of the user. Then use a wine bottle or any clean bottle and press the foil carefully to ensure that everything is stuck well.

7.4.2 Electrolyte Preparation

- **Dissolving the Salt**: Dissolve one tablespoon of salt (NaCl) or baking soda ($NaHCO_3$) in one cup (250 mL) of deionized water.

- **Stirring the Solution**: Mix thoroughly until the salt or baking soda is entirely dissolved, to form a homogenous conductive electrolyte solution.

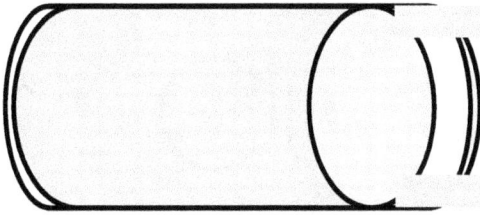

FIGURE 7.3
Rolled-up supercapacitor assembly.

7.4.3 Supercapacitor Assembly

- **Preparing the Separator**: Excise a piece of filter paper slightly larger than the area of the electrodes.
- **Layering the Components**: Meticulously stack the components in the subsequent order: Electrode 1 (carbon-coated side facing upward) → Electrolyte-soaked separator → Electrode 2 (carbon-coated side facing downward). Ensure the separator completely isolates the two electrodes.
- **Securing the Assembly**: Gently compress the layers together to guarantee optimal contact. Roll the foil and the separator tightly to form a cylinder and secure the rolled-up capacitor with tape and insert it in the plastic box or container of cylindrical shape. Avoid applying excessive pressure, which could impair the separator or electrodes. Figure 7.3 shows the rolled-up supercapacitor with the extended aluminum foil for clipping the wires.

7.4.4 Lead Connection

- **Attaching the Leads**: Fix the wire to the extended aluminum foil and ensure a secure and dependable electrical connection.
- **Marking the Terminals**: Explicitly designate the positive and negative terminals to forestall incorrect connections during charging and discharging. It doesn't matter which side you pick; however, once you have chosen the positive and negative, maintain the terminals for both charge and discharge test.

7.4.5 Supercapacitor Testing

- **Initial Voltage Measurement**: Use a multimeter to gauge the initial voltage across the supercapacitor terminals. It should approximate zero.

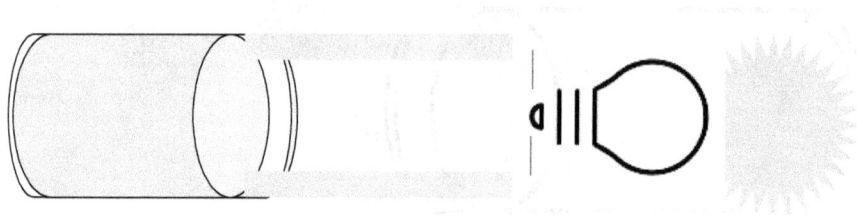

FIGURE 7.4
Supercapacitor with connected LED light.

- **Charging the Supercapacitor**: Connect the supercapacitor to a low-voltage DC power source (e.g., a battery or a regulated power supply) for a few seconds. The voltage should not transcend the safe operating voltage of the components utilized. A current-limiting resistor is recommended to avert damage during charging.

- **Charge Retention Test**: Disconnect the power source and observe the voltage reading on the multimeter. The voltage should remain relatively stable, indicating charge retention.

- **Discharge Test**: Connect a small LED to the supercapacitor terminals to observe the discharge of stored energy. The LED should illuminate briefly, gradually dimming as the supercapacitor discharges. A resistor can be used in series with the LED to regulate the discharge rate. Figure 7.4 shows the supercapacitor connected to the LED light. Table 7.1 shows a template of the experimental readings, which the students are expected to fill it. Also, the students should describe their observation in this experiment.

TABLE 7.1

Experimental Reading Table for the Students to Complete

Time (seconds)	Voltage (V) Charging	Voltage (V) Discharging	Observations
0	0.00	—	
1	1.20	—	
2	2.35	—	
3	2.85	—	
4	3.00	2.95	
5	—	2.60	
6	—	2.30	
7	—	1.95	
8	—	1.60	

7.5 Safety Precautions

- While this method employs safe household materials, it is right to exercise caution when manipulating electricity.
- Refrain from overcharging the supercapacitor, as this could culminate in damage or even fire.
- Do not short-circuit the supercapacitor terminals, as this can unleash a substantial quantity of energy expeditiously.
- Dispose of the supercapacitor responsibly.
- For handling the chemicals, always wear your gloves to avoid irritation and skin burn, especially for strong alkaline solutions.

7.6 Further Exploration

This guide provides a basic introduction to building a supercapacitor. Experimenting with different materials, electrolytes, and designs can improve performance and deepen understanding of energy storage. Using advanced electrode materials like graphene and carbon nanotubes, trying various separators, and testing different electrolytes can enhance the supercapacitor's properties. Additionally, refining assembly techniques and packaging can increase durability and lifespan. This project lays the groundwork for further exploration in electrochemistry and energy storage.

7.6.1 Answer the Following Questions

1. What is the voltage across the supercapacitor before and after charging?
2. How fast does the voltage decay after disconnecting the power source?
3. Is the voltage-time behavior consistent with the theoretical RC time constant?
4. Does the measured capacitance (using voltage and charge) match the rated value?
5. Are there any signs of internal resistance (IR drop) when discharging rapidly?

References

1. D.R. Lobato-Peralta, E. Duque-Brito, H.O. Orugba, D.M. Arias, A.K. Cuentas-Gallegos, J.A. Okolie and P.U. Okoye, "Sponge-like nanoporous activated carbon from corn husk as a sustainable and highly stable supercapacitor electrode for energy storage," *Diamond and Related Material*, vol. 138, 2023, https://doi.org/10.1016/j.diamond.2023.110176.
2. K.C.S. Lakshmi and B. Vedhanarayanan, "High-performance supercapacitors: A comprehensive review on paradigm shift of conventional energy storage devices," *Batteries*, 2023, https://doi.org/10.3390/batteries9040202.
3. Y. Gao, C. Liu, Y. Jiang, Y. Zhang, Y. Wei, G. Zhao, R. Liu, Y. Liu, G. Shi and G. Wang, "Hydrothermal assisting biomass into a porous active carbon for high-performance supercapacitors," *Diamond and Related Material*, vol. 148, 2024, https://doi.org/10.1016/j.diamond.2024.111487.
4. S. Chen, W. Wang, X. Zhang and X. Wang, "High-performance supercapacitors based on graphene/activated carbon hybrid electrodes prepared via dry processing," *Batteries*, vol. 10, 2024, https://doi.org/10.3390/batteries10060195.
5. F. Niaz, S.S Shah, K. Hayat, M.A. Aziz, G. Liu, Y. Iqbal and M. Oyama, "Utilizing rubber plant leaf petioles derived activated carbon for high-performance supercapacitor electrodes," *Industrial Crops and Products*, vol. 219, 2024, https://doi.org/10.1016/j.indcrop.2024.119161.

8

Household Pathogens Removal

Summary

Pathogens including bacteria, virus, and fungi are disease carriers. Therefore, it is important to understand how they work and various ways of inactivation. A simple experimental procedure that could be performed at home is used to demonstrate the effectiveness of different household cleaning methods in reducing microbial contamination on common household surfaces. This hands-on experiment aims to help learners observe the impact of cleaning practices in real time and understand how microbes respond to various cleaning agents typically used in homes. By simulating basic microbiological testing in a home setting, the activity reinforces the importance of hygiene and offers practical skills in scientific observation and experimentation.

8.1 What Are Pathogens

Pathogens are microscopic organisms that cause disease in humans, animals, and even plants. They come in several major types including bacteria, viruses, fungi, and parasites, and they differ widely in their structure and behavior. What connects them all is their ability to invade a host and disrupt normal biological functions, often leading to infection or illness. These organisms are found nearly everywhere: in the soil beneath our feet, the water we drink, the air we breathe, and the food we eat. Despite their size, pathogens have an enormous impact on public health, agriculture, and ecosystems.

Bacteria are single-celled organisms that reproduce rapidly and can live in a wide range of environments. Some bacteria are harmless or even beneficial just like those in our gut. On the contrary, but pathogenic bacteria such as *Salmonella* or *E. coli* can cause severe illness if ingested. Viruses, however, are much smaller and not technically alive; they require a host cell to replicate. When viruses like Influenza or COVID-19 infect a person, they hijack the host's cells and use them to make copies of themselves, damaging tissue in the process.

Fungi include molds, yeasts, and mushrooms. While many are harmless or helpful, pathogenic fungi such as Candida can infect the skin, mouth, or bloodstream, particularly in people with weakened immune systems. Parasites are typically more complex organisms and can range from single-celled proto-zoa to multicellular worms. A notorious example is Plasmodium, the parasite that causes malaria and is spread by mosquitoes.

Because of their differences, pathogens interact with their hosts and environments in unique ways. This has significant implications for how we diagnose, treat, and prevent the diseases they cause. For example, antibiotics are effective against bacteria but not viruses, while antifungals target fungal cell structures that don't exist in bacteria or human cells. Understanding these distinctions is essential for choosing the right cleaning or treatment method.

What makes some microorganisms pathogenic while others are harmless or beneficial?

_____

_____

_____

_____

_____

_____

How do different pathogens enter the body, and why might this matter when designing methods to remove them from surfaces?

_____

_____

_____

_____

_____

Why do you think some pathogens are easier to eliminate than others in certain environments?

_____

_____

_____

_____

_____

8.2 How Cleaning Products Work and Why They Destroy Pathogens

Cleaning products are essential tools for maintaining hygiene and preventing the spread of disease. While basic cleaning removes dirt, grime, and some microbes, disinfection and sanitization are specifically aimed at killing or inactivating pathogens. These products are formulated with active ingredients that chemically disrupt essential parts of microbial cells, such as their membranes, enzymes, or genetic material. This damage prevents the pathogen from reproducing or functioning, rendering it harmless.

Different disinfectants work in different ways. For example, bleach (sodium hypochlorite) is a powerful oxidizer that breaks down cell walls and internal components. Alcohol-based cleaners, such as those with ethanol or isopropyl alcohol, dissolve the lipid membranes of viruses and denature proteins in bacteria and fungi. Hydrogen peroxide releases reactive oxygen species that cause oxidative damage to microbial cells. Other compounds, like quaternary ammonium compounds (quats), which are commonly used in household and hospitals can disrupt membranes of bacteria.

However, the effectiveness of a cleaning product depends on more than just its ingredients. Contact time—the amount of time the surface remains wet with the product—is critical. Many disinfectants require a few minutes to work properly. The type of surface, the amount of organic material present, and the specific pathogen being targeted all influence results. If a disinfectant is wiped away too quickly or used improperly, it may not fully eliminate harmful microbes.

In recent years, there has been growing concern about overusing harsh disinfectants, especially in everyday settings. While powerful chemicals may be necessary in hospitals, using them excessively in the home can lead to unintended consequences such as chemical exposure or environmental harm. Additionally, some microbes may develop resistance to disinfectants, much like they do to antibiotics. This makes it increasingly important to choose cleaning methods that balance effectiveness with safety and sustainability.

Why might a disinfectant that works well on one type of surface or microbe be ineffective on another?

_____

_____

_____

_____

_____

_____

How do you think we can decide whether a cleaning method is "too harsh" for a given environment?

_____

_____

_____

_____

_____

In what ways can we use cleaning science to help prevent the development of microbial resistance?

_____

_____

_____

_____

_____

_____

8.3 How Agar Plates Were Developed and How They Work

Agar plates are an essential tool in microbiology that allow scientists to grow and study microorganisms in a controlled environment. The base of these plates is agar, a gelatinous substance derived from red algae. In the 1880s, microbiologists needed a substance that remained solid at human body temperatures and wouldn't degrade easily. Fannie Hesse, working alongside her husband in Robert Koch's lab, suggested using agar, which she had used in cooking. Its stability and transparency made it ideal for observing microbial colonies.

When agar is mixed with nutrients—such as sugars, proteins, and salts—it becomes a rich environment where microbes can grow. Once a sample is applied to the plate (usually with a sterile swab or inoculating loop), the plate is incubated at a warm temperature to encourage growth. Over time, individual microbes multiply into visible colonies, each one usually arising from a single original cell. The appearance of these colonies can provide clues about what kinds of microbes are present based on shape, size, color, and texture. Agar plates are incredibly useful for comparing microbial contamination before and after cleaning. By swabbing a surface, transferring the sample to a plate, and incubating it, one can visually observe the amount and types of microbial growth. Fewer or smaller colonies after cleaning indicate a more effective disinfection process.

These plates also help researchers isolate specific microorganisms for further study. For instance, if multiple microbes are present on a surface, their colonies will usually grow separately, allowing scientists to test each one's response to different disinfectants or antibiotics. This isolation process is key to identifying pathogens in clinical settings or evaluating hygiene interventions in households or schools.

Why is it important that agar plates provide a controlled and consistent growth environment?

_____

_____

_____

_____

_____

_____

How might different nutrient formulations in agar affect what microbes grow and how well they grow?

_____

_____

_____

_____

_____

_____

What are the limitations of using agar plates to assess microbial contamination, and how could these be addressed?

_____

_____

_____

_____

_____

8.4 Objective of the Experiment

The objective of this chapter is to demonstrate, through a simple at-home laboratory experiment, the effectiveness of different household cleaning methods in reducing microbial contamination on common household surfaces. This hands-on experiment aims to help learners observe the impact of cleaning practices in real time and understand how microbes respond to various cleaning agents typically used in homes. By simulating basic microbiological testing in a home setting, the activity reinforces the importance of hygiene and offers practical skills in scientific observation and experimentation.

In this experiment, students or participants will collect microbial samples from frequently touched surfaces such as kitchen counters, door handles, bathroom sinks, or mobile phones, both before and after cleaning. They will then use accessible materials—such as agar plates (or DIY alternatives), cotton swabs, and common household cleaners—to test and compare the reduction in microbial growth. Cleaning methods may include chemical disinfectants (e.g., bleach, alcohol-based products), natural options (e.g., vinegar, lemon juice), and simple physical cleaning (e.g., soap and water or wiping with a damp cloth).

This experiment not only introduces participants to basic microbiology techniques such as swabbing, culturing, and observing microbial colonies, but also encourages scientific thinking and data collection in a familiar environment. Through visual and measurable results, the chapter highlights which methods are most effective and why, helping participants make informed choices about cleaning practices in their daily lives. The activity also fosters a greater appreciation for the invisible world of microbes and the role of personal responsibility in maintaining a clean and healthy living space.

8.4.1 Pre-Experiment Questions

What are microbes, and where can they be found in the home?

Microbes are tiny living organisms such as bacteria, fungi, and viruses. They are found everywhere—in the air, on surfaces, in water, and even on our skin. In homes, they commonly exist on kitchen counters, bathroom sinks, doorknobs, and electronic devices.

Why is it important to remove microbes from surfaces?

While not all microbes are harmful, some can cause illness or contribute to poor hygiene. Regular cleaning helps reduce the risk of spreading harmful bacteria and viruses, especially in areas where food is prepared or where many people touch the same surface.

What household cleaners will you test in this experiment? Why were they chosen?

Examples may include bleach, vinegar, dish soap, or alcohol-based sanitizers. These are chosen because they are commonly available and have different chemical properties that may influence how well they remove or kill microbes.

What types of surfaces will you test, and why?

Common surfaces like countertops, doorknobs, light switches, or smartphone screens might be selected because they are frequently touched and more likely to harbor microbes.

What do you predict will happen after cleaning each surface?

I predict that chemical disinfectants like bleach or alcohol will show the greatest reduction in microbial growth, while natural cleaners like vinegar may be moderately effective. Water alone might remove some microbes but not as effectively.

What materials and safety precautions do you need for this experiment?

Materials may include agar plates (or DIY gelatin alternatives), cotton swabs, gloves, cleaning agents, and labels. Safety precautions include wearing gloves, avoiding contact with growing microbial cultures, and disposing of samples safely after the experiment.

How will you measure the effectiveness of each cleaning method?

I will compare the number and size of microbial colonies on the agar plates before and after cleaning. A surface with fewer or no colonies after cleaning is considered more effectively disinfected.

What variables will you control to make the experiment fair?

To ensure fairness, I will swab the same area size, use the same type of agar, apply each cleaner for the same amount of time, and incubate all samples under similar conditions.

8.4.2 Materials

- Cotton swabs
- Sterile water
- Petri dishes with agar (store-bought or homemade gelatin agar). These serve as the medium where microbes grow.
- Permanent marker
- **Cleaning Supplies:** Rubbing alcohol (70%), Bleach solution (diluted 1:10 with water), Vinegar, Soap, and water
- Gloves
- Ruler
- Incubator (or warm place like the top of a refrigerator). An actual incubator is ideal, but for home experiments, a warm, dark area like the top of a refrigerator, a cupboard, or a cardboard box with a small lamp inside can act as a DIY incubator.

8.5 Safety Precautions

1. **Wear Personal Protective Equipment (PPE):** Always wear gloves, a lab coat or apron, and safety goggles when handling cleaning solutions, agar plates, and swabs to avoid contact with bacteria or harsh chemicals.

2. **Work in a Clean Environment:** Disinfect your workspace before and after the experiment to minimize contamination and reduce the risk of spreading bacteria.

3. **Handle Agar Plates with Care:** Always hold Petri dishes from the sides and avoid opening them more than necessary. Exposure to open air can introduce airborne pathogens.

4. **Avoid Inhaling Vapors:** When using alcohol, bleach, or vinegar, make sure the area is well-ventilated. Do not inhale the fumes directly, as they can be irritating to your respiratory system.

5. **Label Chemicals Clearly:** Clearly label all cleaning solutions and store them safely when not in use. Never mix cleaning agents (e.g., bleach and vinegar), as this can produce toxic gases.

6. **Dispose of Materials Properly:** After incubation, do not open the plates. Seal them in a plastic bag and dispose of them according to your school or local biohazard disposal guidelines.

7. **Wash Hands Thoroughly:** After completing the experiment, remove gloves and wash your hands thoroughly with soap and warm water to prevent any bacterial transfer.

8.6 Procedure

1. **Label the Petri Dishes**
   - Divide each petri dish into five sections using a marker.
   - Label them as follows: **Control, Soap & Water, Alcohol, Bleach, Vinegar** (Figures 8.1 and 8.2)

2. **Clean the Surface:**
   - Wipe the surface using different cleaning methods: (1) **Soap and water,** (2) **Alcohol,** (3) **Bleach solution,** (4) **Vinegar** (Figure 8.3)

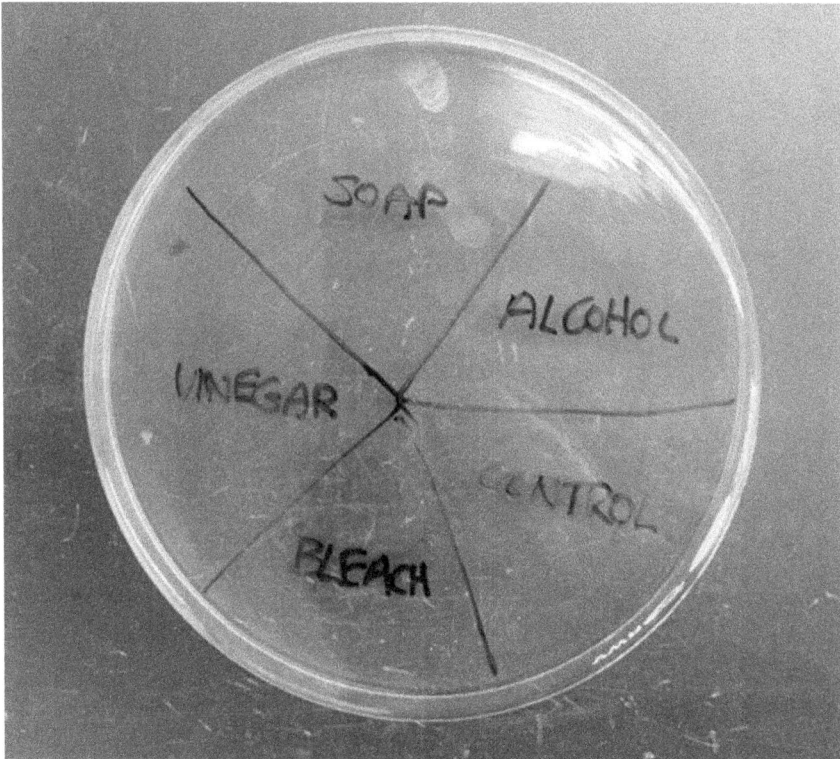

FIGURE 8.1
Petri dish labeled in five sections. Make sure to clean the petri dish before handing, handle it with gloves, and to write on the BACK of the disk. Writing on the front of the dish can cause contamination that will skew your experimental results.

3. Collect Control Samples (Before Cleaning):

- Choose a common surface (e.g., kitchen counter, doorknob, bathroom sink).
- Use a sterile cotton swab dampened with sterile water to wipe the surface.
- Gently streak the swab onto each section of the agar plate. Use a new swab for each cleaning method and streak the corresponding sections on the agar plate (Figure 8.4).

4. Incubate the Plates:

- Store the plates in a warm, dark place (~30°C–37°C if possible) for 24–48 hours (Figure 8.5).

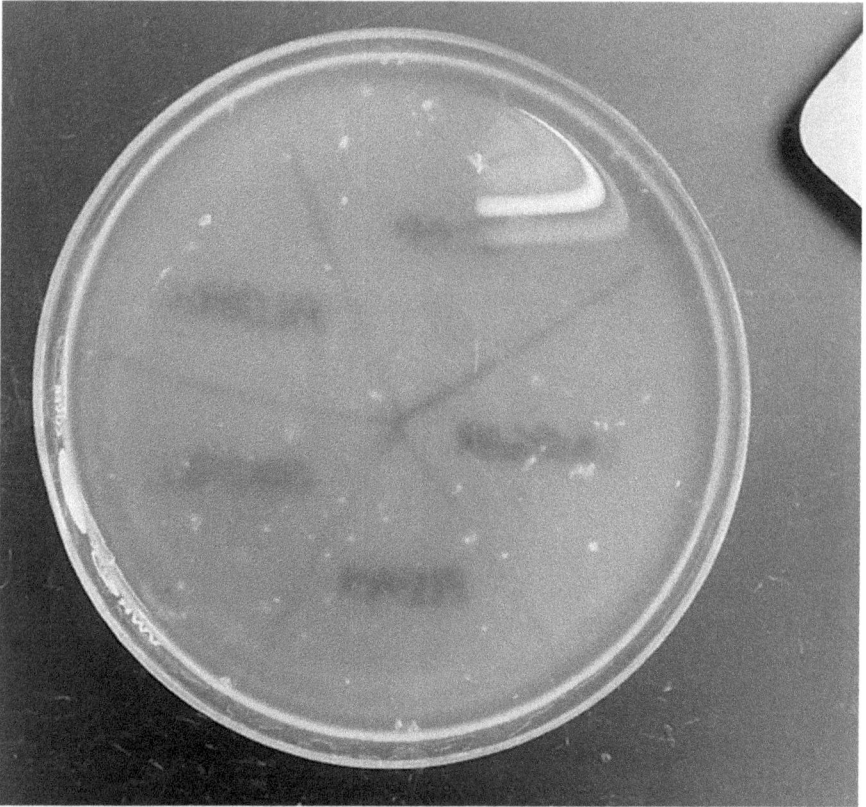

FIGURE 8.2
If you are starting with a petri dish without agar, then add in your agar solution and allow for the solution to become a solid. Make sure that this process is done under as sterile of conditions as possible (use gloves, clean the plate). While the plate is curing, try to have it sit in a place that does not have open air flow or with a lid on top to minimize the number of pathogens blown onto the plate.

8.7 Observations and Results

To properly read an agar plate, begin by examining the bacterial colonies that have grown on its surface. These colonies typically appear as distinct, circular formations that vary in size, color, and texture. Larger colonies often indicate faster-growing bacteria, while differences in color and shape can help distinguish between bacterial species. Some colonies may appear smooth and shiny, while others may have a rough or fuzzy texture, especially if mold is present. Carefully noting these characteristics is essential in assessing bacterial growth and contamination levels.

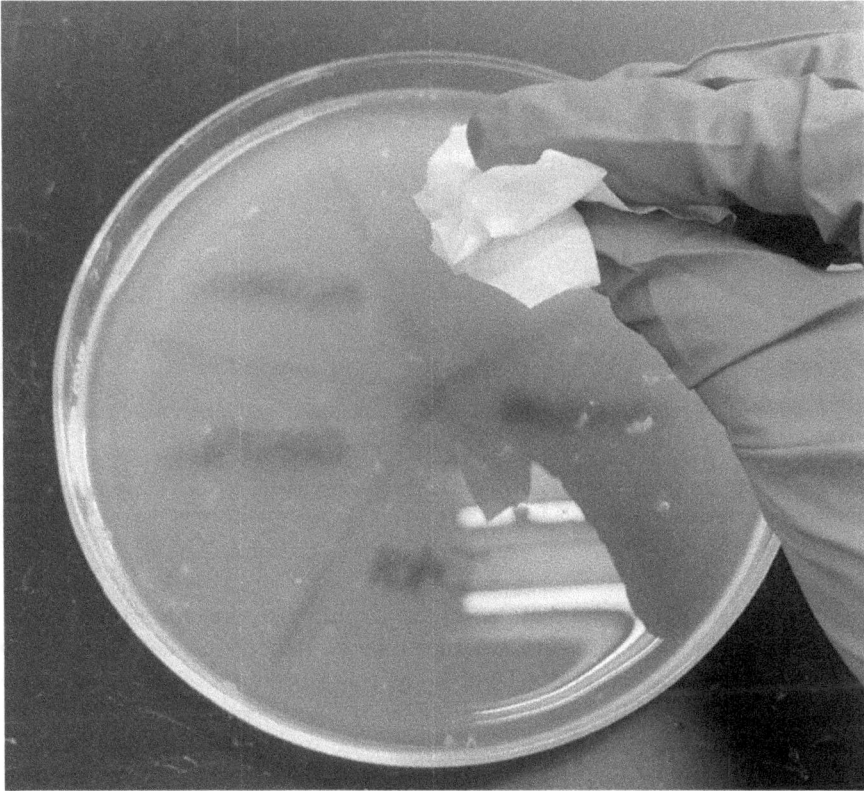

FIGURE 8.3
Gently wipe each surface with its respective cleaner.

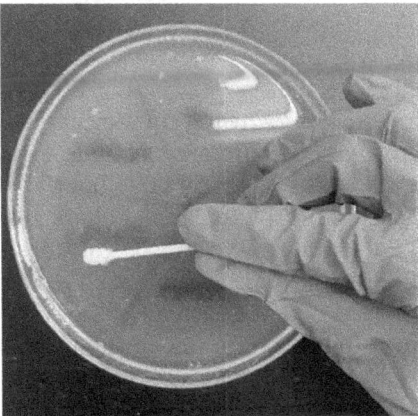

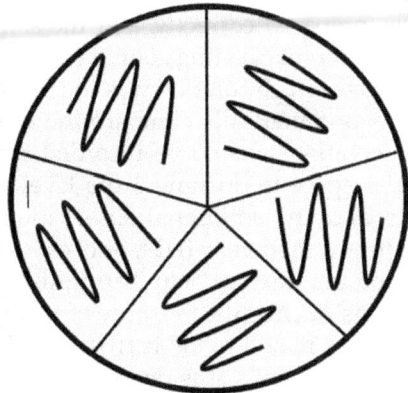

FIGURE 8.4
Collect the sample and then gently streak the sample onto the agar plate. Try to avoid disturbing the top layer of the plate. Make sure to wear glove.

FIGURE 8.5
Set your sample in an area that is warm and dark, for example the top of a refrigerator. Try to find an area where the agar plate will receive minimal air flow to prevent the disturbance of the plate.

The dirtiest areas of the plate will have the highest concentration of bacterial growth. In these sections, colonies may be densely packed together or even merge into a continuous layer known as a bacterial "lawn." If an agar plate has been swabbed from multiple sources, the section with the most abundant and largest colonies represents the most contaminated area. In addition to bacteria, the presence of fuzzy or discolored growth may indicate mold, which suggests further contamination.

Conversely, the cleanest areas of the plate will have little to no visible bacterial growth. If only a few small, isolated colonies appear, it suggests that the surface swabbed was relatively clean. If a section of the plate remains entirely clear, this could indicate that no bacteria were present, provided that the agar was prepared correctly, and the sample was properly handled. When testing different cleaning solutions, the plate with the least bacterial growth can indicate which solution was the most effective in reducing contamination.

To systematically compare the effectiveness of different cleaning agents, observations should be recorded for each agar plate. The table below provides space to document the level of bacterial growth, colony characteristics, and any additional observations. Bacterial growth can be categorized as heavy, moderate, light, or none, while colony characteristics such as color, shape, and texture help distinguish different types of bacteria. Other observations, such as the presence of mold, unusual smells, or unexpected growth patterns, may provide further insight into contamination levels. By analyzing these results, it is possible to determine which cleaning agent was most effective in reducing bacterial presence (Table 8.1).

TABLE 8.1

Results Documentation Table

Sample Type	Growth Ranking	Bacterial Growth Description	Colony Characteristics	Additional Results or Comments
Control				
Bleach				
Vinegar				
Soap and water				
Alcohol				

Which cleaning agent was most effective in reducing bacterial growth? Why do you think it was more effective than the others?

_____

_____

_____

_____

_____

_____

How did the appearance of colonies differ between the control plate and the treated plates? What does this tell you about the effectiveness of the cleaning agents?

_____

_____

_____

_____

_____

_____

Were there any unexpected results, such as mold growth or unusual colony shapes? What might have caused these irregularities?

_____

_____

_____

_____

_____

_____

How do you think the concentration or application method of the cleaning agents could affect their effectiveness?

_____

_____

_____

_____

_____

_____

Did any of the plates show minimal or no growth at all? What factors could explain this outcome?

_____

_____

_____

_____

_____

_____

How do the colors and textures of bacterial colonies provide insight into the types of bacteria present? Were there any surprising colony characteristics?

_____

_____

_____

_____

_____

_____

What role do environmental factors (e.g., temperature, humidity) play in bacterial growth on the agar plates?

_____

_____

_____

_____

_____

Based on the results, what are the advantages and disadvantages of using natural cleaning agents (like vinegar and soap) versus chemical disinfectants (like bleach and alcohol)?

_____

_____

_____

_____

_____

If you repeated this experiment, would you change any of the methods or variables (e.g., type of agar, application technique)? How and why?

_____

_____

_____

_____

_____

How do the results of this experiment relate to real-world scenarios, such as sanitation in public spaces or food preparation areas?

_____

_____

_____

_____

_____

_____

Index

Note: **Bold** page numbers refer to tables and *italic* page numbers refer to figures.

For Product Safety Concerns and Information please contact our EU
representative GPSR@taylorandfrancis.com
Taylor & Francis Verlag GmbH, Kaufingerstraße 24, 80331 München, Germany